"Business Guides on the Go" presents cutting-edge insights from practice on particular topics within the fields of business, management, and finance. Written by practitioners and experts in a concise and accessible form the series provides professionals with a general understanding and a first practical approach to latest developments in business strategy, leadership, operations, HR management, innovation and technology management, marketing or digitalization. Students of business administration or management will also benefit from these practical guides for their future occupation/careers.

These Guides suit the needs of today's fast reader.

Business Guides on the Go

Mario Vanhoucke

A Quest for Projects with Scarce Resources

Seeking Schedule Intelligence Through Project Data Discovery

Mario Vanhoucke
Faculty of Economics and Business Administration
Ghent University
Gent, Belgium

ISSN 2731-4758　　　　　　ISSN 2731-4766　(electronic)
Business Guides on the Go
ISBN 978-3-031-71506-8　　　ISBN 978-3-031-71507-5　(eBook)
https://doi.org/10.1007/978-3-031-71507-5

© The Editor(s) (if applicable) and The Author(s), under exclusive license to Springer Nature Switzerland AG 2024

This work is subject to copyright. All rights are solely and exclusively licensed by the Publisher, whether the whole or part of the material is concerned, specifically the rights of translation, reprinting, reuse of illustrations, recitation, broadcasting, reproduction on microfilms or in any other physical way, and transmission or information storage and retrieval, electronic adaptation, computer software, or by similar or dissimilar methodology now known or hereafter developed.
The use of general descriptive names, registered names, trademarks, service marks, etc. in this publication does not imply, even in the absence of a specific statement, that such names are exempt from the relevant protective laws and regulations and therefore free for general use.
The publisher, the authors and the editors are safe to assume that the advice and information in this book are believed to be true and accurate at the date of publication. Neither the publisher nor the authors or the editors give a warranty, expressed or implied, with respect to the material contained herein or for any errors or omissions that may have been made. The publisher remains neutral with regard to jurisdictional claims in published maps and institutional affiliations.

This Springer imprint is published by the registered company Springer Nature Switzerland AG
The registered company address is: Gewerbestrasse 11, 6330 Cham, Switzerland

If disposing of this product, please recycle the paper.

If your dreams do not scare you, they are not big enough.
                                             Ellen Johnson Sirleaf

Preface

This book tells the story of two researchers who joined forces to find better project data to improve their academic challenges for project scheduling with resources.

Our journey began with a serendipitous encounter—a chat between two conference attendees through lunch break triggered insights that sparked new ideas and perspectives. This encounter and the unexpected access to a powerful computer marked the beginning of a transformative chapter in our lives. Armed with newfound inspiration and computational capabilities, we tackled the challenge of solving one of the most challenging project scheduling problems in the academic literature. It became a journey combining the rainy weather of Belgium with beautiful sunny days in the capital of Portugal. It also became a story of perseverance and friendship.

In this book, we offer a synthesis of our findings—a summary that describes the essence of our research efforts published in various studies. We invite you to join us on this discovery as we navigate through our academic exploration guided by curiosity, passion, and the relentless pursuit of answers to our questions.

Let us unravel the mysteries of the unknown together to illuminate the path toward new horizons of understanding.

Welcome aboard, academics and practitioners, fellow seekers of knowledge in managing projects more effectively.

The journey awaits.

Ghent, Belgium and Lisbon, Portugal Mario Vanhoucke
July 2024

Introduction

Welcome to the fascinating world of academic research, where every breakthrough raises new questions, initiating a never-ending quest for knowledge. In this book, we will embark on a journey through the exploration of using project data to solve the well-known resource-constrained project scheduling problem (RCPSP).

The problem of scheduling projects with limited resources is recognized as a central optimization problem within the domain of project management and scheduling. It is known as a highly complex challenge that researchers and practitioners have tackled, leading to extensive research and analysis within the academic community.

At the core of the RCPSP lies the efficient allocation of resources to a set of activities within a project, while adhering to various constraints and objectives. These constraints typically include limitations on the availability of resources, such as personnel, equipment, or budgetary constraints, as well as precedence relationships between activities, describing the sequence in which they must be performed. The primary objective in solving the RCPSP is to develop a schedule that minimizes the duration of the project while meeting all these constraints. Achieving this goal requires a challenging exploration of the space of all possible schedules that meet the imposed constraints, with the optimal schedule often hidden among an immense number of other, non-optimal ones.

Over the years, numerous approaches have been proposed to address the RCPSP, ranging from mathematical programming models to heuristic and meta-heuristic techniques. These methodologies attempt to explore the extensive solution space inherent to the problem as effectively as possible, seeking optimal (in the case of exact methods) or near-optimal (in the case of meta-heuristic methods) schedules that meet the project objectives within the given constraints. Despite its inherent complexity, the RCPSP also holds significant practical relevance, with applications spanning various sectors, including construction, manufacturing, and software development. By providing insights into efficient resource allocation and project scheduling strategies, the research on the RCPSP has contributed to improved practices in project management, leading to enhanced productivity, cost-effectiveness, and project outcomes.

With the continued and growing attention to this problem within the field of project management, there is ongoing progress and increasing understanding of the complexity of this problem and the development of new and efficient scheduling methodologies to solve it. This book does not aim to provide a comprehensive overview of the current state of academic research but rather offers a glimpse into the fascinating journey we undertook to better understand this problem. It is a story of the quest for high-quality project data for the problem. A story that began with an encounter with a fellow researcher that has evolved into a journey through project data and a quest to unravel the complexity of the problem by meticulously analyzing the intricate details of project data. This journey involved transforming projects, making them more complex or simpler, all with the overarching goal of improving our knowledge and understanding of the complexity of the problem.

It became a journey that turned a simple encounter into an intense friendship, that transformed a visit to Lisbon into a passionate love for this beautiful city, and went from a MacBook laptop to a powerful computer that could condense years of experiments into just a few precious research weeks.

Without these serendipitous life-changing events, this book would never have been written.

Two Life-Changing Events

The quest began with a short 10-minute meeting with a colleague (José Coelho) at a project scheduling workshop in Valencia (Spain), where we engaged in a brief conversation and arranged a potential collaboration that—at best—would lead to one publication. His insights and our shared passion and drive to explore the unknown led us to embark on a lifelong quest without even realizing it. This initial encounter in 2002 became a life-changing event (the first of two) that enabled me to write this book. Everything you will read from here on is based on years of research that we published in different esteemed academic journals. It has been an intense search, with the unavoidable ups and downs and the blissful feeling when a new publication appears, but it all started with this short and insignificant first encounter.

Without this encounter, we could never have embarked on this journey, but the real breakthrough of our research came many years later, when José had already become a friend, and our research ideas became increasingly challenging and intense.

With the growing number of projects in our database, the need for more experiments and powerful computers also grew, to the point where our laptops, however powerful, were no longer suitable for our ambitions. Due to my frequent visits to Lisbon (where José is a professor) and his visits to Ghent (where I work), it seemed appropriate to appoint José as a guest professor at our university so that we could collaborate even more effectively. Once that was arranged, José enthusiastically set up the supercomputer infrastructure at Ghent University, which I did not know how to use. With his strong computer expertise and boundless enthusiasm for things that most people would not get excited about,[1] the availability of this supercomputer became the driving force behind our ambitions to subject the growing database of available projects to an almost infinite series of computational tests.

---

[1] I remember José passionately discussing the number of cores the supercomputer had and how fast the memory could be used on a beautiful evening at the restaurant with our partners. I saw the wonderful looks from our partners toward us; it seemed like pure love, but it could also have been pity because they did not understand what was being discussed.

José's appointment became an entrance ticket to immense computing power and changed our research lives for a second time. From that moment on, we could tackle our challenging project scheduling problem with strength, determination and almost no limitations.

The Problem

Since this book focuses on the quest for data for the RCPSP, the reader must have a good understanding of this well-investigated project scheduling problem. There are indeed many versions of this problem available (some of which are discussed in Chap. 10), but the basic version is the one that academics have investigated the most.

A project is represented by a so-called activity-on-the-node network, where the nodes represent activities and the arrows between the nodes represent precedence relationships (which determine that one activity can only start after its predecessor has finished). Various types of precedence relationships are possible, but in the basic version of the RCPSP, so-called *finish-to-start* relationships are used with a time-lag of zero (indicating that each activity can start immediately after all of its predecessors have finished). To create a schedule for such a project, the simple and well-known *critical path method* can be used, which is a very efficient and straightforward way to determine the earliest possible end time of the project.

However, this method does not consider the limited availability of resources.

The RCPSP assumes the presence of renewable resources for the execution of activities. Renewable resources are available on a period-by-period basis, i.e., the available amount is renewed from period to period (e.g., per hour or day), and thus, the total resource use at every instant is constrained. Typical examples include manpower, machines, tools, equipment, and space. Each activity uses a certain number of these renewable resources, referred to as resource demand or resource requirements, and this demand must, of course, be less than the number of available units of these resources. The RCPSP also operates under a so-called *fixed duration* mode, meaning that the duration cannot be changed.

For example, an activity with a duration of 3 days and a resource demand of 4 units requires these 4 units each day during its execution. In some practical settings, however, this fixed duration may not always be the case, as 3 days with 4 resources can be represented as 12 person-days, leading to possible increases (or decreases) in the duration by assigning fewer (or more) resources to it. However, such changes are not possible in the basic version of the RCPSP. As for the availability of renewable resource types, they are assumed to be known and fixed (in terms of units), meaning that they will never change over the life of the project. In practice, this may (again) not be true, as it is often the case that the availability of skilled laborers may vary over time due to absence, sickness, or planned vacations. However, in the basic version of the RCPSP, this is not possible at all.

The RCPSP is a well-known and extensively researched project scheduling problem, but it follows very strict assumptions.

The goal of the RCPSP is to create a schedule so that each activity has a start time (and an end time) that respects both the precedence relationships (from the network) and the limited availability of all resources. Creating such a schedule is not very difficult, but the challenge lies in the fact that it should have the smallest possible duration. This total duration, often referred to as the project makespan, is the objective of this problem and makes it one of the most difficult scheduling problems in the project management literature (mathematicians refer to it as an NP-hard problem). Since the number of possible schedules for such a problem is huge, the challenge is finding one (or possibly more than one) among all these possible schedules that minimizes the makespan. When mathematicians conclude that this search is an NP-hard search, it means that it is often a *mission impossible* for realistic projects.

No wonder academics have come to love this problem so much.

Figure 1 displays an example project network with 11 activities and finish-start precedence relations between them. Each number above the node denotes the estimated duration of the activity, while the number below the node is used to refer to the resource demand (it is assumed here that only one type of resource is available). The availability of the resource is restricted to six units for all periods of the project. The right

Introduction

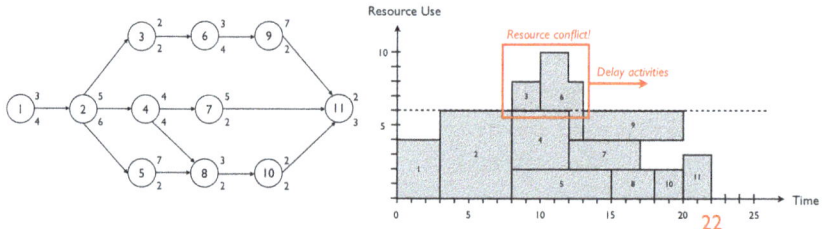

Fig. 1 Solving resource conflicts

part of the figure displays an earliest start schedule, obtained by simple critical path calculations that respect the precedence relations of the project but do not consider the resource constraints. Each activity is represented by a rectangle, with the horizontal length representing the activity duration and the vertical height representing the renewable resource requirement/demand. The earliest start schedule shows a resource over-allocation between periods 8 and 13 since the total use of the renewable resource by activities 3, 4, 5, 6, and 7 exceeds its limited availability of 6. Such over-allocation is called a *resource conflict*.

When a resource conflict arises in a schedule, researchers see a challenge flashing before their eyes.

Solving the RCPSP is therefore a search for the best possible way to resolve these resource conflicts so that the project makespan is minimized. It is at this point that the complexity of the problem emerges. Solving resource conflicts entails shifting certain activities to reduce the resource usage, ensuring it does not exceed the resource availability, and choosing which activities to shift is not easy. In the example, activities 3, 4, and 5 are scheduled at time 8, and thus, multiple combinations of shifts are possible (each of which can lead to a different schedule). Moreover, each shift might cause a new resource conflict later in the schedule, requiring another choice to be made. This sequence of choices and shifts results in a cascade of resource conflicts until all of them will eventually disappear, resulting in a so-called feasible project schedule with a certain duration (the word "feasible" is used to indicate that all constraints are met). When a feasible schedule has the smallest possible makespan, it is called an optimal solution to the problem. The search for such optimal solution(s)

Introduction   xv

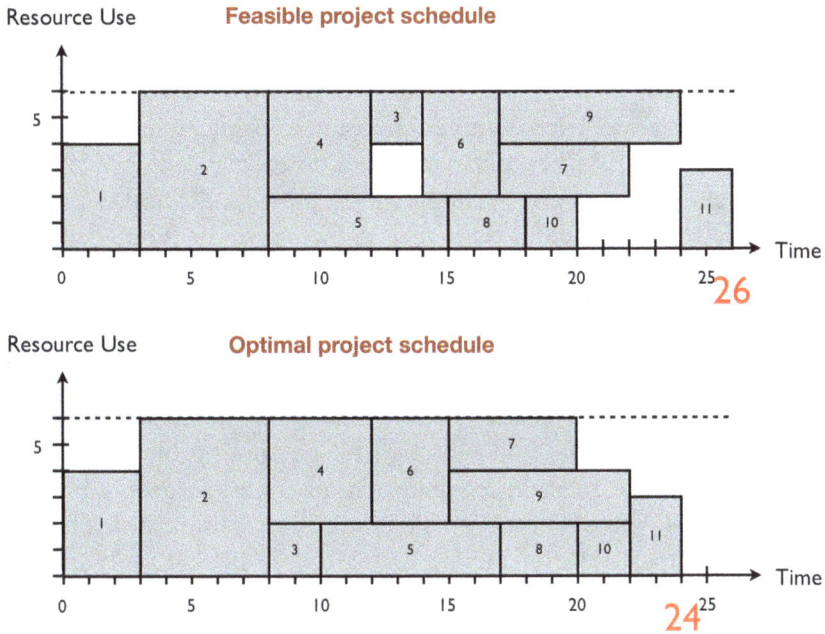

Fig. 2 Feasible and optimal schedule

can be like finding a needle in a haystack, as the number of choices (and schedules) can be so large that even the most powerful computers cannot enumerate them all.

In the top schedule of the Fig. 2, activities 3 and 6 were chosen to be shifted so that the over-allocation of the resource (the resource conflict) disappeared. A simple shift of these activities resulted in a resource-feasible project schedule with a total project makespan of 26 time units. When a different choice of activities would have been made, for example, activities 5 and 6, it would lead to a different resource-feasible schedule. With this better choice, a project schedule makespan of only 24 days can be obtained, as shown in the bottom schedule. This schedule has the lowest possible project duration and is thus the optimal project schedule.

The reader who is new to this domain should be aware that the example project is only an illustrative example of a problem that is rather easy to solve. Larger projects with more activities (and therefore larger networks) and with multiple resource types (rather than one type as in the example) are, of course, the projects for which the problem becomes a real challenge. This book describes the quest for projects that are challenging to solve. When we have complex projects available, it enables us to understand precisely what makes the problem so challenging. However, the book also explores ways to simplify these projects. If complex projects can be made easier, there is hope that we can solve them one day. This will provide us with information and insights that simplify the search for a robust and efficient solution algorithm.

So, we set out to find both difficult and easy projects for the RCPSP. Perhaps we are simply rather demanding.

This quest for high-quality project data brings together a series of studies presented in the book through ten interesting but challenging research questions. You will see that each question has a sort of beauty that sounds like music to the ears. Their melody has unparalleled beauty, but there is also a form of hidden dramatization behind every musical line. That is precisely what has made this journey so fantastic. The best music is the one where melody and melodrama work together.

Ten Studies

The research into the secrets of the complexity of the RCPSP is obviously not my only passion in life. I am also very addicted to music (preferably on vinyl), with a preference for progressive rock albums that have a whole narrative in the lyrics. Their music is often ingeniously crafted like a complex spiderweb through which the listener must maneuver to understand (and appreciate) the whole story. It is quite common in progressive rock albums for songs to have the same title (with their differences indicated by different parts). This approach allows artists to create an extended musical narrative and a sequence across multiple tracks. It is a way to explore a theme or concept in depth while providing a cohesive listening experience for the audience. The "Part 1" and "Part 2" labeling helps to signify that these songs are connected in some way, whether musically, thematically, or narratively. It is a technique that many

Fig. 3 New album (out now!)

progressive rock bands have used to add depth and continuity to their albums.

I wish I could write such progressive albums too.

However, I cannot imagine a better job than being an academic researcher. Luckily, the process of conducting academic research and presenting the results in workshops is not so different from the creative process of recording an album and promoting it to a larger audience on stage (I think).

That is why I would like to present the ten research questions addressed in this book as 10 different songs (chapters) on a vinyl record with a side A and B. Despite the importance of an attractive album cover for vinyl records, I kept it simple yet beautiful, as you can see in Fig. 3. Of course, as with all good music, the album is made with guest musicians making significant contributions to most songs. I could never write music on my own.

Imagine this: While the audience is waiting to listen to the vinyl album for the very first time on a high-quality stereo deck (no Spotify, only analog listening allowed), the following announcement is read aloud on the radio to build anticipation:

Introducing the electrifying new project data-driven progressive rock album by Mario Vanhoucke, featuring collaborations with the renowned and famous artist José Coelho and other artists from his research team. This dynamic album, spanning two sides and boasting a total of 10 captivating songs, takes listeners on an unforgettable musical journey. From intricate melodies to soaring guitar solos and intense computer experiments, each track is a testament to Vanhoucke's and Coelho's innovative vision and the collective talent of the collaborating artists. Prepare to be swept away by the raw energy and emotive storytelling woven throughout this genre-defying masterpiece. Their latest album "A quest for projects with scarce resources" is a must-have addition to any progressive rock aficionado's collection.

Here is the track listing of our new album.

Side A. The Quest for Project Data (featuring José Coelho)

Chapter 1. Do We Have Enough Project Data? (Part 1)
Chapter 2. Do the Projects Exist in Reality?
Chapter 3. Do We Have Good Schedules for the Projects?
Chapter 4. Can We Solve Every Project Instance?
Chapter 5. Why Is the RCPSP So Difficult? (Part 1)
Chapter 6. Can We Make the Projects Easier? (Part 1)
Chapter 7. Why Is the RCPSP So Difficult? (Part 2)
Chapter 8. Can We Make the Projects Easier? (Part 2)

Side B. The Relevance of Project Data (featuring Other Researchers)

Chapter 9. How to Make the Project Data Practically Relevant?
Chapter 10. Do We Have Enough Project Data? (Part 2)

The search for answers to these ten research questions became a story of trial and error, with every new insight triggering even more new questions. The scheduling problem became a passion and challenge that has added color and joy to our academic careers. True research is only done when you navigate your way to the deepest secret places of a problem. Just as music begins where words end, our quest began anew each time our previous research question was answered.

After all, everything that lives has a rhythm.

Contents

Part I Side A. The Quest for Project Data (Featuring José Coelho)   1

1 Do We Have Enough Project Data? (Part 1)   5
    1.1 Life-Changing Event #1   6
    1.2 Generating Projects   8
    1.3 Predictive Power   13

2 Do the Projects Exist in Reality?   19
    2.1 Project Classification   20
    2.2 Empirical Data   24
    2.3 Just a Beginning   28

3 Do We Have Good Schedules for the Projects?   31
    3.1 Where Is the Data?   32
    3.2 More Is Better (*Sometimes*)   35
    3.3 No More Data   38

Contents

4 Can We Solve Every Project Instance? — 41
  4.1 Life-Changing Event #2 — 42
  4.2 From 12 to 48 Algorithms — 44
  4.3 Solving Instances to Optimality — 49
  4.4 Hunger for More — 52

5 Why Is the RCPSP So Difficult? (Part 1) — 55
  5.1 Small Projects Are Easy — 56
  5.2 Going to the Core — 58
  5.3 44 Years on a Computer — 62
  5.4 Small Projects Are Not Easy — 66

6 Can We Make the Projects Easier? (Part 1) — 69
  6.1 Solution Equivalency — 71
  6.2 Changing Resources — 74
  6.3 Reliable Resource Indicator — 78
  6.4 Resource Strength — 82

7 Why Is the RCPSP So Difficult? (Part 2) — 85
  7.1 It Is No Luck — 87
  7.2 In Distance Lies Complexity — 89
  7.3 Going to the Core (*Again*) — 93
  7.4 New Dataset (*Again*) — 97

8 Can We Make the Projects Easier? (Part 2) — 99
  8.1 Solution Equivalency (*Extended*) — 101
  8.2 Cover Sets — 104
  8.3 Reducing the Search Space — 107
  8.4 Four Times More Data — 111
  8.5 Easier Projects — 112
  8.6 Mission Accomplished — 117

Part II Side B. The Relevance of Project Data (Featuring Other Researchers)		119
9	**How to Make the Project Data Practically Relevant?**	121
	9.1 Calibrating Data	124
	9.2 Empirical Validation	129
	9.3 Human Partitioning	130
	9.4 Human and Statistical Power	132
	9.5 More Empirical Data	134
10	**Do We Have Enough Project Data? (Part 2)**	137
	10.1 Resources	138
	10.2 Flexibility	141
	10.3 People Skills	143
	10.4 Project Portfolios	145
	10.5 Where Is Lisbon?	146
Afterword		147
	Why I Write Books	148
	The Future of Project Data	151
Research Background (*a bit tedious to read*)		153
	Side A. The Quest for Project Data (Featuring José Coelho)	153
	Side B. The Relevance of Project Data (Featuring Other Researchers)	156
	References	157

About the Author

Dr. Mario Vanhoucke is a Professor at Ghent University (Belgium), Vlerick Business School (Belgium), and UCL School of Management at University College London (UK). He teaches courses on Project Management, Applied Operations Research, and Decision-making for Business. His research interests lie in the integration of project scheduling, risk management, and project control, which has led to more than 100 papers in international journals, five project management books published by Springer, one book published by Apress, and a PM Knowledge Center for online learning. His research has received multiple awards, e.g., from the Project Management Institute (PMI Belgium), the College of Performance Management (CPM), and the International Project Management Association (IPMA).

Part I

Side A. The Quest for Project Data (Featuring José Coelho)

The quest for good project data is as old as the research in project management and scheduling itself. No researcher enjoys working in a completely theoretical environment where thought experiments are sufficient to draw conclusions. Instead, most of us are researchers who seek solution methods in the form of optimization algorithms, simulation models, and analytical or statistical methods to search for ways to better manage and schedule a project. In the specialized field of project scheduling with resources, the vast majority of researchers intensely study the principles of operations research and aim to construct a feasible schedule through a mathematical model programmed in one programming language or another. The main objective of such research is to find better schedules than those previously discovered in prior studies, allowing us to publish our new algorithms in reputable professional journals.

For such research, we need access to project instances.

We need data. We want projects.

Fortunately, they are also sufficiently available.

The most well-known dataset in project scheduling is the so-called project scheduling problem library (abbreviated as PSPLIB) proposed in a paper in 1996 by Rainer Kolisch and Arno Sprecher. It was a time when algorithms for solving the RCPSP began to work better than ever, and where the available projects at that time (known as the Patterson set) gradually became too easy. There was even a procedure available[1] that could optimally solve all 110 project instances from these Patterson instances. That was a breakthrough at that time, of course! But the fun was gone, leaving no reason for any researcher to continue investigating this challenging problem. Yet, the RCPSP was not entirely solved; it was just that the used Patterson projects were too small, and therefore, too easy to solve. With the increasing computing power and the growing strength of the algorithms, researchers were thus forced to seek out more difficult and challenging project instances, and that is where the authors of the PSPLIB cleverly intervened.

Not that they just came up with random projects, as that would never have had the impact that the PSPLIB has had. Instead, these two researchers realized that a set of projects must be carefully composed so that a number of important parameters, likely determining the complexity of the projects, are included in the generation of this data. And so, the authors decided to generate the projects under a controlled design, ensuring that such complexity parameters are indeed varied in these projects. The popularity of PSPLIB is, therefore, attributed to its diverse set of problem instances that cover various aspects of project scheduling, allowing for comprehensive testing and analysis. These instances help researchers understand how well their algorithms perform under different constraints and conditions.

The specific contribution of the generated dataset is that the projects differ in size (ranging from 30 to 120 activity projects) and in the structure of their network (ranging from serial to parallel projects), but also that it contains projects with different values for the resource constraints. This last aspect was achieved by varying the values of a metric now known as the resource strength (RS), ensuring that the set ranges from easy to difficult instances, and everything in between. In hindsight, it seems like a logical choice to vary the network and resource structure, but at the time, it was the great insight and contribution of the PSPLIB, and the reason why these 2040 projects received so much attention. To this day, the

---

[1] That was the branch-and-bound procedure by Erik Demeulemeester and Willy Herroelen, published in Management Science in 1992. These authors were my supervisor and co-supervisor respectively, and you can see that I inherited the love for the RCPSP from them.

Side A. The Quest for Project Data (Featuring José Coelho)

PSPLIB remains the most used dataset to test new algorithms in project scheduling research, and as we will see later in this book, this is entirely justified.

Yet, we should not be blind to some possible shortcomings. Despite the set's great utility, we must be willing to question the importance of the set, if only because our research spirit requires it (never take anything for granted, and keep seeking improvements).

After all these years, there is perhaps a need for a new dataset, just as there was during the time of the Patterson projects. It is not the case that all project instances can already be optimally solved,[2] as was the case with the Patterson projects, but that does not automatically mean that these projects are still useful for further research.

Perhaps the dataset is no longer diverse enough, and maybe the resource strength is no longer good enough to describe the project complexity.

Perhaps the research has evolved to the point where the current algorithms will not gain much additional insights and potential improvements if we continue to test on the same project data.

Who knows. Who will say. We had so many questions.

Precisely these questions, along with our curiosity about the complexity of scheduling projects with resources, prompted José and I to spend a few years of our lives searching for new, and hopefully better, project data.

And so we started our collaboration.

"It is a good reason to visit Lisbon for a while," I must have thought, but I never knew I would end up staying there.

---

[2] At the time of publication of this book, 1666 of the instances were optimally solved (closed instances), and for the remaining 374, we did not know whether their best-known solutions could be improved or not (open instances).

1

Do We Have Enough Project Data? (Part 1)

Note: This chapter is based on the article "*An evaluation of the adequacy of project network generators with systematically sampled networks*", published in *European Journal of Operational Research*.

The search for the importance of high-quality project data already began long before the PSPLIB was introduced in the literature. Perhaps there is no clear starting point in this challenging data quest, but a study from 1980 by Saleh Elmaghraby and Willy Herroelen might be one of the most important milestones that still resonates in current academic research. What they claimed back then still has an impact to this day, as they wrote the following:

> A choice between two proposed algorithms, or the determination of the efficiency of a particular algorithm, would be greatly facilitated if there exists a measure of network complexity. This would eliminate any possible bias in the conclusions regarding the efficiency of a particular algorithm relative to others by ensuring that the algorithm is evaluated at several points in the range of complexity.

With this argument, the authors mainly emphasized the importance of being able to measure the structure of the project network with an indicator (which they called a network complexity measure) to distinguish between easy and difficult projects.

I wrote in the prologue of Part I that the PSPLIB was the first project dataset generated under a controlled design, and that this was an important milestone in understanding the difficulty of projects. During the generation of these projects, not only the network structure (using a network indicator) but also the resource constraints (using a resource indicator) were very carefully controlled. However, such design does not necessarily mean that the entire spectrum of possible projects, representing the full range of complexity, is automatically covered. To ascertain this, we need to get an idea of how large the space of all possible projects is, and that is a challenging task given the wide diversity of possible networks.

But nothing is impossible.

With a fast project network generator, sufficient computing power, and a method to measure the dimension of that network space, this could potentially be investigated. I had no clue how to start, but a new research idea was born.

And it thoroughly shook up my work life.

1.1 Life-Changing Event #1

I immediately got to work on this new idea and dusted off my old random network generator (RanGen1[1]) to generate new networks. I was not very satisfied with the performance of this generator and was working on a second version of it (RanGen2) to measure the network structure of the generated networks in a new, hopefully better, way. In the RanGen1 version, I used the so-called order strength as the network indicator to measure how close the project network lies to a fully serial or parallel network, but I had seen that there existed other network indicators

---

[1] Despite the fact that the study on the RanGen1 network generator was only published in 2003, I already had it at my disposal by the end of 1999, and I knew already in 2002 that the network generator could be improved.

proposed in a Portuguese study with the meaningless names I_2 to I_6. I wanted to incorporate these new network indicators into my RanGen2 version, but there was an error in one of the formulas, so I could not proceed with the programming. I could contact the authors of course, asking them for help, but they might never reply, so I waited patiently. After all, there was a conference scheduled not long after, where I might possibly meet one of them.

It was April 2002, and I had only graduated from my PhD one year prior, so it was my very first conference without my supervisor by my side.[2] That eighth Project Management and Scheduling workshop that I attended as a young assistant professor in Valencia (Spain) became one of the most important conferences that shaped my future career. It was there that I spoke with José for the very first time. Our conversation did not last long, just a brief moment, and it went something like this:

> **Mario:** Hi José, nice to meet you. I am working on implementing your I_2 to I_6 indicators, but I think there is an error in the formulas. Could you take a look at what is wrong?
> **José:** Interesting, I'll check on this. It's probably a typo in the publication. I don't have the original files from the research with me right now, but I'll look into the formula once I'm back home in Lisbon.
> **Mario:** Lisbon, that's Portugal, right? Never been there, but thanks! Let's keep in touch. I'll explain later what I'm exactly working on.
> **José:** Sure, we stay in touch.

The rest of the workshop was undoubtedly interesting, although I cannot remember much more of it. I did not have much hope for the "*we stay in touch*" part either, as that is often said at conferences, and usually nothing comes of it. But this time was different. José replied immediately upon returning home with a correction of a minor error in his publication, and he also provided some comments and suggestions to

---

[2] It is not that my supervisor had turned his back on me, but I had left the University of Leuven to start as an assistant professor at Ghent University. I did not even have a budget to attend conferences at that time, and Erik was kind enough to pay for my trip one more time, even though I no longer worked for him.

improve the research idea right away. Best of all, I could now continue programming my RanGen2 generator, and I started right away.

It was not so much the speed of José's reaction that surprised me, but rather the fact that there was someone else in the world who found project network generations as interesting as I did, and that seemed unique and an opportunity I could not let pass. I explained to him that I wanted to generate as many project networks as possible with my new network generator. I also told him that I not only wanted to use my RanGen2 generator but also the well-known ProGen generator that was used to generate the PSPLIB networks. I explained that comparing these two network generators seemed like an interesting idea, even though my main goal was to generate as many different networks as possible to determine if they cover the full range of complexity.

José responded to my emails with great enthusiasm and immediately offered his own network generator (called RiskNet), so ultimately, we had three completely different project network generators available. With these three network generators, we thought it should be possible to cover the entire space of projects as comprehensively as possible. And that is exactly what we tried to find out right away.

For those who have not noticed yet, I wrote a lot in the first-person singular form up to now. I was at that time indeed completely on my own in a new university looking for new research challenges. But from the moment that I met José in Spain, everything started to change rapidly. From that moment on, we began our search for project data together. No more "I," but "we."

1.2 Generating Projects

With our three network generators (ProGen, RanGen, and RiskNet), José's five network indicators (I_2 to I_6), and a sufficiently powerful computer, we got to work and tried to generate as many project networks as possible with a fixed size of 30 activities (similar to the J30 set of the PSPLIB). To more or less delineate the space of all possible networks, we divided the possible values of the five indicators into different parts, as shown in Fig. 1.1. Some years after our first study, we decided to give

Indicator (Original name)	Indicator (Current name)	Meaning	Values (Tested)
I_2	SP	Serial/Parallel	0 to 1 (steps of 1/29)
I_3	AD	Activity distribution	0 to 1 (steps of 0.01)
I_4	LA	Length of short arcs	
I_5	-	Length of long arcs	
I_6	TF	Topological float	

Fig. 1.1 Network indicators for projects

these network indicators a different, more accessible name, as you can see in the figure.

The whole idea of splitting the possible values of these network indicators into smaller parts was to describe the entire space of project networks. Indeed, each network has a value for each of these indicators, and by partitioning these indicators into parts, we could find out in which part of the space a particular network belonged.

The figure shows that the I_2 indicator, which we later called the "serial/parallel indicator," consists of 30 possible values for 30-activity networks (ranging from 0 to 1 in steps of 1/29). The other indicators I_3 to I_6 always range between 0 and 1 with more than 30 possible values, and so we decided to divide them into 100 classes in steps of 0.01. Therefore, the total search space was described in a matrix of $30 \times 101^4 = 3{,}121{,}812{,}030$ cells, which were initially all empty. Each time a new project network was found during the generation, the values for the five indicators were calculated, and their position in this five-dimensional space was determined. Then, the network was added to this particular cell to indicate that at least one network exists with this particular combination of I_2 to I_6 values. By generating many networks using the generators, we attempted to fill as many empty cells as possible to understand the true size of the project space.

It might sound like an easy search, but it was not.

In most programming languages, an integer is represented using 32 bits (although it can now also be 64 bits), which means we needed 11.62 gigabytes of memory to create this matrix. For younger readers, this may seem like a challenging but not impossible task, but with the computers of the time, it was simply a mission impossible. However, since we only wanted to report whether or not there exists a project for each cell, and we were not particularly interested in how many networks existed in each cell, we represented each cell with one bit (binary programming) instead of an integer, reducing the necessary space to 416 MB of RAM. Initially, each cell was set to 0, and from the moment the generator found a network, that value was set to 1, providing us with the necessary information. Peanuts with today's computers, but a challenge at the time of this research study.

Our laptops generated networks for multiple days in a row to set as many cells to 1. We had to restart the search several times to adjust the parameters of the network generators and steer them in a different direction in the search space, but we were determined to carry out this quest as well as possible. While the computers tested our patience, we comforted ourselves with the expression that "waiting is not mere empty hoping. It has the inner certainty of reaching the goal".[3] But our patience was eventually rewarded.

What we saw after many days even exceeded our expectations. In total, exactly 19,105,294 cells were set to 1. This may seem disappointing at first (as it covers only 0.61% of our maximum allowed search space), but we had no reason for any disappointment. The maximum space of 3,121,812,030 cells really is of little significance and does not at all indicate that this is actually the real space where projects exist. There are likely many combinations of the network indicators (cells in the five-dimensional matrix) for which projects do not exist at all, and so this huge matrix with cells does not provide any information about the theoretical maximum number of possible projects. No, 0.61% was not bad at all.

---

[3]This quote comes from the *Book of Changes* (known as "the I Ching"), and it is highly relevant and applicable to academic research.

1 Do We Have Enough Project Data? (Part 1)

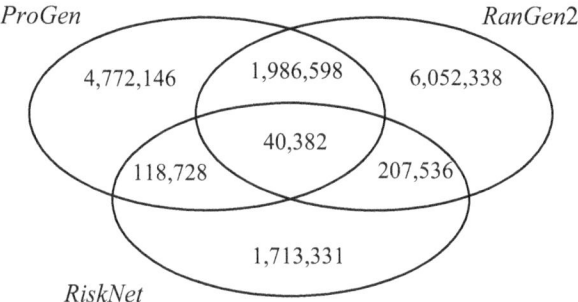

Fig. 1.2 Generated networks (cells)

What struck us most was that none of the three network generators worked perfectly. Each generator filled the space in a different way, as can be seen in Fig. 1.2. Only 40,382 types of networks could be found by all three generators, but many cells could be set to one by one generator, while the other generators could not find any network there. For example, the figure shows that a total of 4,772,146 networks[4] were found by ProGen that could not be generated by any other generators. Moreover, up to 7,973,205 networks (6,052,338+207,536+1,713,331) were found by the others that ProGen could never find. Given that ProGen was used in generating the PSPLIB networks, it might suggest that the networks in this set lack diversity, and many other networks that could have been included are absent from it.

Of course, our intention was not to criticize the PSPLIB set and portray it as worthless (because it is not). The projects from the PSPLIB were generated under several constraints (such as allowing maximum three successors for each activity), which narrows the space of generated projects. In our experiments, no such restrictions were imposed as we wanted to obtain as many different networks as possible. So comparing the results of our experiments with the PSPLIB projects is not entirely fair, but we still want to make that comparison. However, we could not

---

[4] Actually, I should write *types of networks*, as the figure shows a 1 if at least one network was found in that cell.

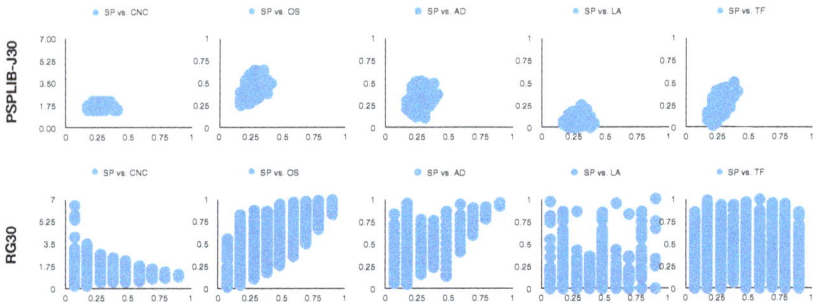

Fig. 1.3 The network structure for the PSPLIB-J30 and RG30 sets

help but conclude that the projects from the PSPLIB set occupy only a very small portion of the space of all possible projects.

What this experiment taught us is that it is indeed possible to better fill the space of all networks, which led us to decide to create a new dataset that is more diverse than the PSPLIB set. However, to not deviate too much from the PSPLIB philosophy, we kept the number of activities to 30 (just like the J30 set) but varied the other I_2 to I_6 indicators in such a way that we obtained a wide variety of new projects. Each time, we varied the value of one indicator between its minimum and maximum values, while the values for the other indicators were set randomly. For example, we generated networks for the SP indicator between 0 and 1 in small steps to cover the entire SP space while letting the other indicators vary without any control. We did this for each indicator separately, so that we ultimately ended up with five different sets. In total, we obtained 1800 new projects generated with the new RanGen2 generator, and we called this set the RG30 set.[5]

Figure 1.3 compares the network structure for the J30 instances of the PSPLIB set (top row) with the structure of the RG30 networks (bottom row). Each graph compares a known network indicator on the y-axis with

---

[5] We also varied the values for different resource indicators, but these were not studied in detail in our study, and they will be extensively discussed in later chapters of the book.

the serial/parallel (SP) indicator on the x-axis.[6] The coefficient of network complexity (CNC) indicator is the network indicator used to generate the PSPLIB instances, and the order strength (OS) serves as an alternative to the SP indicator. The other indicators, AD, LA, and TF, are three indicators from Fig. 1.1. The graphs clearly show that the diversity in the network structure has significantly increased for the new RG30 instances.

And so with this new set, we gained access to new and different projects that we could use for our research. We were strongly convinced that with the 3840 available projects (2040 from the PSPLIB and 1800 from the new RG30 set), we would be much better equipped to solve the RCPSP. After all, we could conduct many more tests on much more diverse projects, which we believed would only benefit further research.

But just because we were convinced of it does not mean it is true, and the question remained whether more diverse projects can indeed add value to academic research.

1.3 Predictive Power

The truth is that we had no ambition to answer this question ourselves, and we decided to leave it to others in their future research. Time would tell, we thought. But we could not resist giving a little hint to illustrate the utility of multiple project datasets. You must remember from the quote at the start of this chapter that the initial purpose of the study was not to create a new dataset of projects (although we did this with the RG30 set) but rather to investigate whether the networks adequately covered the full range of complexity.

Therefore, we found it useful to conduct our own experiment with the new RG30 projects to see if the greater diversity in projects could also lead to better insights when using (existing or new) algorithms to schedule them. We argued previously that both network and resource indicators describing a project should be able to distinguish between

---

[6] If you are wondering why the SP indicator receives such central focus, it is because many studies have shown that this indicator is one of the most important for measuring the network structure of a project.

easy and difficult projects. If we could predict this difference using these indicators, we could much better choose the right algorithm to create the best schedule for the project, thus solving the RCPSP more effectively. Therefore, in our experiment, we were investigating whether the RG30 projects, which have a much more diverse network and resource structure than the existing projects up to that date, could detect such a distinction between easy and difficult projects. In a study from 1999 by Willy Herroelen and Bert De Reyck, this transition between easy and difficult projects was examined through phase transitions.

A phase transition in computer complexity refers to a significant change in the behavior of a computational problem as certain parameters are varied. These transitions often mark shifts in the computational difficulty of solving the problem. In the context of algorithmic analysis, phase transitions can indicate critical points where the problem becomes significantly harder or easier to solve, or where the properties of the solutions change abruptly. Understanding phase transitions is important in algorithm design and complexity theory as it helps in identifying the boundaries of tractable and intractable problem instances.

In order to find whether the network indicators could reveal such phase transitions, we used one of the best performing algorithms from the literature to solve the RCPSP to optimality. More specifically, we used the branch-and-bound procedure developed by Erik Demeulemeester (my advisor during my doctoral studies), known for its ability to quickly produce optimal schedules when the projects are not too large. This branch-and-bound method consists of a powerful and super-efficient search for optimal solutions, dividing the project into smaller parts (branching) performed in a tree structure. Some branches of the tree need not be further explored because it can be demonstrated that the best schedule could not be found (dominance rule), while other branches must be further split until a feasible schedule is found. The algorithm continues branching and searching until all nodes created in the tree are explored, eventually resulting in the best feasible schedule. Therefore, the number of created nodes in this tree reflects the difficulty of the project because each node requires a number of calculations, and with a growing number of these nodes, the computer time can quickly escalate from seconds to minutes, hours, days, and even more.

Fortunately, many of the RG30 project instances were solvable within a reasonable time,[7] and so we could easily determine the number of created nodes for each project after our computer runs. With a dataset of 1800 projects, each with known values for the network indicators and the number of created nodes in the branch-and-bound tree, we could now examine whether there was indeed a link between the indicators and the complexity of solving the problem. Today, we would apply a number of machine learning algorithms to find this out, but the study was conducted at a time when artificial intelligence and machine learning were not as popular yet. Therefore, we decided to use a CHAID regression.

CHAID (Chi-squared Automatic Interaction Detection) is a statistical technique used for creating regression trees, which are predictive models that recursively partition the project data into subsets based on the predictor variables. In our case, the predictor variables are the values of the network indicators, as we aim to use them to predict the complexity of the scheduling problem, referred to as the target variable, which is measured by the number of nodes created in the branch-and-bound tree. The CHAID algorithm starts with the entire dataset and identifies the indicator variable (from I_2 to I_6) that best splits the data into two or more homogeneous groups based on the value of the target variable (the number of created nodes). At each step, the algorithm evaluates the significance of the split (using chi-squared tests) to maximize the homogeneity of the obtained groups. The goal is for the final end nodes to contain projects with known values for the indicators, where the complexity of the projects in this group is approximately the same (and significantly different from the projects at other end nodes of the CHAID tree).

Figure 1.4 shows the split of the RG30 set, ultimately displaying 21 different end nodes for which the values of the I_2 to I_6 indicators are known. For example, end node 1 represents a group of projects with $I_2 < 0.08$ and $I_3 < 0.49$ values. This end node only uses two indicators, but for other end nodes, almost all indicators were necessary to define

---

[7] However, some of them could not be optimally solved at all, which came as a sort of surprise to us, as will be discussed in Chap. 4.

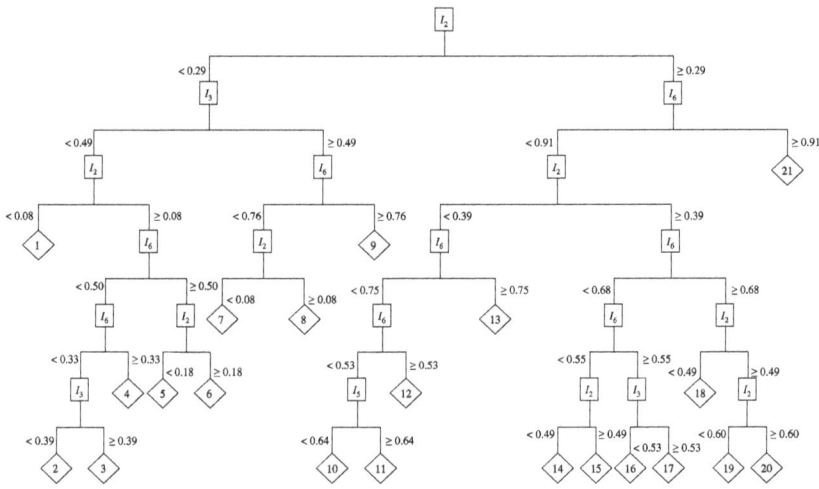

Fig. 1.4 CHAID regression tree (RG30)

the group of projects with identical complexity. It is noteworthy that the I_4 indicator was never used in the CHAID search, and the I_5 indicator, which is very similar to I_4, was only used once, meaning that these indicators have no strong discriminatory value.[8]

For each new project, the values of the network indicators can be easily calculated, allowing us to determine which end node the project belongs to. This, in turn, enables us to predict how difficult it will be to achieve an optimal solution for this project using the branch-and-bound procedure (and perhaps other procedures as well). So, if you know the characteristics of the project (indicators), you also know how difficult it will be to find a very good schedule. Powerful, isn't it?

But is what we claim here also correct?

To verify this, we generated 200 new projects that we had never used before, each with different values for the network indicators. These

---

[8] Because the I_4 and I_5 indicators closely resemble each other, we decided to come up with a more pleasant name for only one of the two, just as we did for the other indicators (see Fig. 1.1). Since the I_4 indicator is easier to understand than the I_5 indicator, we ultimately preferred renaming the I_4 indicator.

1 Do We Have Enough Project Data? (Part 1)

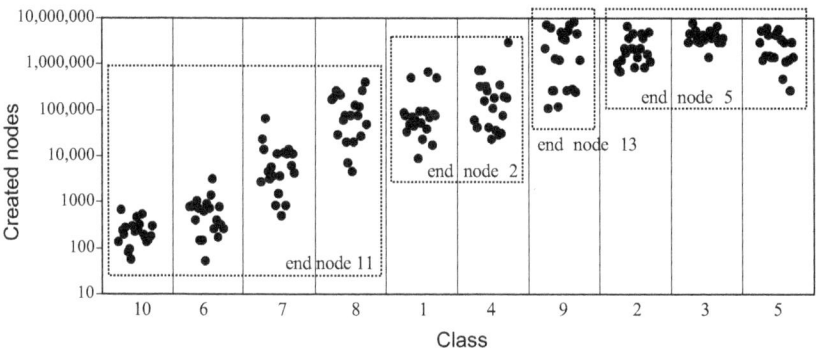

Fig. 1.5 Predictive power (new projects)

projects were created using our RanGen generator across 10 different classes, with each class containing 20 projects. We then examined the end nodes of the CHAID tree to determine their locations. Because the generation of the 200 new projects happened more or less randomly, we obtained projects from only a few end nodes (end nodes 2, 5, 11, and 13 to be precise). We proceeded to solve these 200 projects optimally using the branch-and-bound procedure and observed the time it took, measured by the number of created nodes. If our claim was correct, projects from the same class should have approximately the same number of created nodes, while projects from different classes should have a significantly different number of created nodes. To our great delight, what we had hoped to see indeed became a reality.

Figure 1.5 shows that the 10 classes of each 20 projects (x-axis) belonging to four different end nodes of the CHAID tree indeed have a different complexity (y-axis). For example, it is clearly visible that all projects belonging to classes 2, 3, and 5 (with values from end node 5) have a greater complexity than, for instance, the projects from classes 6, 7, 8, and 10 (belonging to end node 11). This suggests that for each new project, where the indicator values are easily calculable to determine its position in the end nodes of the CHAID tree, it is possible to provide a good indication of its difficulty level. When the network indicator values of a project change in such a way that the project ends up in a different end node of the CHAID tree, we are thus dealing with a change in complexity, and therefore with a phase transition.

The much more diverse projects from the RG30 dataset thus have the potential to detect such phase transitions, and ultimately, to better tailor new algorithms for solving the RCPSP to the characteristics of the project.

Of course, these results came only from a relatively small experiment and were just the beginning of a much more intensive search for higher-quality project data (discussed in the next chapters). The results were promising, but naturally, this small experiment also had some limitations. The conclusions were certainly not automatically generalizable to other algorithms, and much more (and better) testing was needed before drawing general conclusions to fully address the phase transition issue. Nevertheless, this study has demonstrated that greater diversity in project data can be useful for further research, and that new algorithms should be tested on a wider variety of projects than what is currently being done.

Ultimately, our very first study was a success, and after several rigorous revisions, it was published in *European Journal of Operational Research* in 2008. I never thought that the brief meeting in Spain in 2002 would yield such a beautiful result. What this study has taught me is that I found this collaboration with José Coelho to be a pleasant experience. At that time, I did not know that this was just the beginning of one of the most intense and rewarding collaborations I would ever experience in my entire career.

But that became very clear to me not much later (*so keep reading*).

2

Do the Projects Exist in Reality?

Note: This chapter is based on two summary articles. The article "*An overview of project data for integrated project management and control*", published in *Journal of Modern Project Management*, provides an overview of the current artificial projects, while the article "*Construction and evaluation framework for a real-life project database*" published in *International Journal of Project Management*, provides details about our empirical project database.

If in the previous chapter I gave the impression that we had "only" 3840 projects at our disposal with the PSPLIB set and the new RG30 set, then I must correct this misunderstanding. There were already many other sets of projects available at that time, and some of them we had generated ourselves in other previous studies.

Moreover, it was not the case that all projects consisted of generated artificial data, as some projects, albeit a minority, came directly from practical applications. And despite the fact that all these projects were intensively used by both José and myself to conduct our research, we were sometimes confused and could not always easily see the forest for the trees.

It was time for an overview.

2.1 Project Classification

At the time of the study in the previous chapter, with the RG30 set, we had made our very first contribution to the generation of project data. However, upon reviewing previous research, we found that it had already been done before, albeit under a less controlled design. Reflecting on this, we discovered that I had actually been generating project data myself for quite some time, without even being aware of it, and without José's knowledge.

To map out the abundance of data, we decided to gather all existing project datasets from my own previous studies. It quickly became apparent that each project dataset had been generated with a specific purpose in mind. Sometimes it was because we needed projects that were larger than the existing PSPLIB projects; other times it was because we wanted to expand the RCPSP to a project scheduling problem with cash flows (and thus needed cash flow data). We decided to view all datasets from the perspective of the RCPSP, ignoring any additional data present (such as cash flows) so that only the basic data remained.

And that basic data for the RCPSP naturally consists of a project network (activities with precedence relations between them) and the resource data to describe the demand and availability of the renewable resources. These two important dimensions, network and resources, form the core of the RCPSP and must be quantified to describe the full range of complexity as we discussed in the previous chapter.

The network indicators were already introduced in the previous chapter, where the I_2 to I_6 indicators were renamed to SP (I_2), AD (I_3), LA (I_4), and TF (I_6).[1] There were, of course, other network indicators known in the literature, among which the order strength (OS) and the coefficient of network complexity (CNC) are perhaps the most well-known. In the

---

[1] Recall that no name was given for I_5 because it seemed too similar to the I_4 indicator, and the CHAID regression did not provide much indication that these two indicators could distinguish between easy and hard projects.

previous chapter, the OS and CNC were not discussed in detail because we did not use those two indicators much in our future research (and they will not be used in the following chapters either). The reason is that the OS is quite similar to the SP indicator, as they both measure the degree of seriality/parallelism of a project. They both have a value between 0 (completely parallel network) and 1 (completely serial network) but are just slightly different in interpretation. Personally, I prefer SP over OS because I can visualize more clearly how a network with SP = 0.5 looks (while for a network with OS = 0.5, this is a bit harder). The reason we omitted the CNC is that quite some studies have shown that this indicator fails to discriminate between easy and hard instances and, therefore, cannot serve as a good measure for describing the impact of network topology on the difficulty of a project scheduling problem. We were convinced that with the four indicators SP, AD, LA, and TF of Fig. 1.1, we could sufficiently describe the network structure of any project.

The reader may have already noticed that in the previous chapter, we mainly focused on network indicators and placed resource indicators somewhat in the background, which was not entirely fair if our love for the RCPSP was genuine. After all, resource indicators are just as important (or perhaps even more important) for describing the complexity of a project, as they reflect the limitations of resources and are ultimately responsible for the so-called resource conflicts that we introduced in the introduction of this book.

The four most well-known resource indicators are shown in Fig. 2.1.[2] The resource use (RU) and the resource factor (RF) both determine how many units the project activities require of each resource type, while the resource strength (RS) and the resource constrainedness (RC) measure the availability of each resource type. As was the case with the network indicators, there are numerous studies that examine these indicators in detail (something we will also do in Chap. 6). And again, sometimes criticism is directed at some of these indicators (as was the case with

---

[2] The formulas for these four indicators can be found in numerous studies and are omitted here for simplicity reasons.

Indicator (Abbreviation)	Indicator (Full)	Meaning
RU	Resource use	Average number of resources used per activity
RF	Resource factor	Average percentage of resource types used per activity
RS	Resource strength	This indicator is a bit more complex and compares the maximum resource use of the critical path schedule with the resource availability and the maximum use of each activity (and since it incorporates a critical path analysis, it is no "pure" resource indicator)
RC	Resource constrainedness	Average number of resources divided by the resource availability

Fig. 2.1 Resource indicators

the CNC). For example, the RS indicator received some criticism for not being a pure resource indicator, as its value depends on the project network and not purely on the resource data.[3] Moreover, while some experiments have shown that the RS works really well to predict the problem complexity, other experiments showed the opposite. There is never a real complete consensus, but researchers seem to be quite adept at criticizing established practices, and perhaps that is how it should be in academia. Despite this criticism, we have used the RS in many of our studies of the following chapters, and you will see that it is not such a bad resource indicator after all.

But as I mentioned earlier, the intention was to collect existing project datasets from previous studies and examine their network structure and resource data based on the indicators discussed earlier. After collecting sufficient data from previous studies, we arrived at the following project datasets:

- The **RG300** dataset originates from a study in 2007 with Dieter Debels where we developed a new and powerful genetic algorithm to solve the RCPSP. In our experiments, we needed large projects, and since the

---

[3]The resource strength calculates the maximum resource use as the peak in usage in an earliest critical path schedule and thus takes the project network data into account.

PSPLIB set only contains projects with a maximum of 120 activities, we generated 480 new instances with 300 activities using the RanGen generator. We were able to demonstrate that our procedure could easily find a near-optimal schedule even for such large projects, and over time, this genetic algorithm became one of the algorithms we frequently used in our future studies, as you will notice in the following chapters.

- The **DC1** dataset originates from a study during my PhD, where I addressed the resource-constrained project scheduling problem with discounted cash flows. This dataset comprises 1800 projects with 10 to 50 activities generated using the ProGen/Max generator, which is an extension of ProGen. Although the dataset also includes cash flows, which are irrelevant for the RCPSP and therefore not utilized further in our studies, the networks and resource constraints remain highly pertinent.
- The **DC2** dataset stems from a subsequent study on the RCPSP with cash flows and includes 720 projects with 25 to 100 activities. Since the RCPSP does not involve cash flows, we excluded them from the projects (similar to what was done for DC1) and retained the remaining networks with resources as a new dataset for our RCPSP research.
- The **MT** dataset is somewhat unique as it was generated for a completely different purpose than outlined in this book. It originates from a study on project control without resource constraints, containing only project networks without resources. The dataset comprises four subsets with 900 + 800 + 1200 + 1200 = 4100 instances. Due to the lack of resources, these projects may seem less relevant for the RCPSP (where resources play a crucial role). However, the dataset proved to be very valuable during the development of a new set some years later, as will be discussed in the next chapter.

The attentive reader has undoubtedly kept track of the total number of projects, knowing that we now have a total of 10,940 projects (comprising the aforementioned RG300, DC1, DC2, and MT projects, as well as the 3840 projects from the PSPLIB and RG30 sets). Many of these played a central role in our future studies, which the following chapters will show. With so many projects, it is easy to lose track, which is why we decided to write our first summary article and also created a website detailing the project data, solutions, and values for all network and resource indicators for each project (you can also download an Excel file for convenience).

At that time, we still did not know if more project data would yield better research results, but we had a plenty of projects available now. In the following chapters, it will be demonstrated that this abundance of data has certainly assisted us greatly in shaping our new ideas.

Perhaps more data is simply better after all.

2.2 Empirical Data

The overview article was not only co-authored by José, but also by Jordy Batselier, who was closely involved in our project data overview for a different reason. Jordy had not been involved at all in the generation of our artificial datasets that we just discussed, but rather focused on the classification of 51 real empirical projects that I had collected over the years.

As a matter of fact, José and I had been wondering for a while whether all those artificially generated projects could actually occur in real life, and the then PhD student Jordy came at the right moment to join our team.

The idea of collecting a set of real projects had started years earlier, born more out of necessity than out of interest in taking a look at the real world. After all, I had started giving project management courses to business people, and they needed more than just the RCPSP. Drafting a resource-feasible schedule was, of course, one of the steps in managing projects, but in addition to that, they also needed a robust risk analysis method (to predict what can go wrong with the project) and a system to monitor the progress of the project (to timely detect the project problems). For that reason, I began to expand my research horizon from resource-constrained project scheduling to risk analysis and project control and named this schedule/risk/control trilogy "Dynamic Scheduling." This expanded research agenda has added color to my academic career and has taken me to places I would never have visited otherwise, but I have written that story in one of my other books.[4]

That is why I had collected quite some projects from companies over the years. They were scattered here and there on my hard disk, but I had never found the courage and time to organize them. That is, until Jordy joined my research team. He turned the jungle of data from 51 projects into a structured framework and presented the very first empirical project dataset to the academic literature (in the 2015 article mentioned at the beginning of this chapter). These empirical projects contain network and resource data to construct a schedule (which is the theme of the current book) but also contain data for risk analysis and project control (which, except in this chapter, are not further addressed).

Structuring the data of empirical projects may seem like a simple and time-consuming task, but it is not as easy as it might seem. Empirical projects fundamentally differ from artificially generated projects because they are not generated by network generators under a controlled design but are obtained through interviews with project managers. Not only does this mean that there are imperfections and errors in the data, we also needed a different way to structure the empirical projects. The values for the network indicators could be easily calculated (because real projects also need a project network); however, for the resource indicators, this was much less the case (because some assumptions about the use of resources were not always met, or because sometimes there simply was no resource data available). Therefore, we opted for a different approach and classified the empirical project data in a completely new way by using two new indicators which we called "completeness" and "authenticity" as shown in Fig. 2.2.

The completeness of the project data is defined as the extent to which each of the three project components (schedule, risk, and control) is covered by the data. This completeness is expressed by a three-level color code based on the traffic light approach. A green indicator signi-

---

[4] As a matter of fact, most of my other books describe the project management research studies from these three dimensions (schedule/risk/control), and in my free book "*The Art of Project Management: A Story about Work and Passion*," I provided an overview of the places I visited and the amazing people I have met thanks to this research.

	Completeness	Authenticity
Schedule	Activity durations and cost Project network Resource data	Project authenticity: The source of the schedule and risk data (coming from project managers, or based on self-made assumptions)
Risk	Standard probability distributions (triangular) or advanced statistical probability distributions	
Control	Real project progress data and EVM performance metrics	Tracking authenticity: The source of the control data (coming from project managers, or based on self-made assumptions)

Fig. 2.2 Empirical classification

fies complete availability of all data. Yellow and orange signals denote moderate and poor data completeness, respectively. For instance, in certain project networks, only activity durations may be available, with partial cost and resource data (yellow). When no cost or resource data is available whatsoever, and only the network data remains, it would turn the completeness into red. However, when the empirical project data is comprehensive, matching the level of information in artificially generated projects, the completeness is set to green.

In addition to indicating whether the project data is complete or not, the quality of the data must also be measured. The concept of authenticity is used to indicate the source of the data and the level of assumptions used in collecting the data. Some parts of the data come directly from the project manager and are therefore fully authentic. For other parts, we had to supplement the data based on self-made assumptions, which obviously makes them less authentic. A distinction was made between project authenticity, which describes the origin of the static data (especially the schedule and also the risk data inputs), while tracking authenticity reflects the dynamic control data measuring the performance of the project during its progress.

And so, we ended up with 51 projects at the time of our publication that were not neatly and accurately generated like the artificial project data to cover the *full span of complexity* but had a clear link to reality.

	Size		Network topology				Schedule	
	# Act	# Res (38%)	SP	AD	LA	TF	Time (Days)	Cost (Euro)
Average	92.4	7.3	41%	57%	15%	36%	279.7	43,389,766
Minimum	7	1	1%	17%	0%	0%	2	1,210
Maximum	1,796	27	95%	100%	100%	100%	2,804	5 billion

Fig. 2.3 Empirical project key statistics

The empirical dataset grew rapidly thereafter to encompass many more projects and is now freely available on our project data website (wait for Chap. 9 for more information on that).

We calculated the values of the network indicators for these projects, shown in Fig. 2.3, and we soon saw that many empirical projects did not overlap with the PSPLIB projects. The number of activities (#Act) ranges from 7 to 1796, bringing much more variation in the project size (the PSPLIB set has a maximum of 120 activities, and the RG300 set contained, up to that point, the largest projects with 300 activities). The number of resources varied between 1 and a maximum of 27, while all artificial projects maximally use four resources. The values for all network indicators vary greatly between their minimum and maximum values, indicating that a diverse artificial dataset with many diverse network indicator values, like the RG30 set, is not unrealistic. It is also worth mentioning that we observed that for 62% of the projects, no resources were available, rendering them irrelevant for the resource-constrained project scheduling studies conducted by José and myself. In fact, constructing a schedule for these projects is very straightforward and can be done using simple critical path calculations.

With more than 10,000 artificial projects and only a few empirical projects, the balance between theory and practice may not seem entirely equal. Nevertheless, this appeared to us to be the ideal starting point to enhance our understanding of the RCPSP.

The main advantage of artificial projects is, of course, that they enable researchers to test their algorithms on a broad and diverse set of projects rather than on a limited set, such as empirical projects. Through this process, researchers can identify which projects pose challenges in finding a good schedule and which ones are relatively easy. Additionally, for empirical projects, a schedule already exists, albeit often not the optimal one. However, despite this, empirical projects can still be valuable for academic research. The existing schedules for these projects are typically created by project managers with the assistance of software tools and could potentially be further improved by academic algorithms. Therefore, assessing the quality of existing real schedules to determine whether academic algorithms can enhance them can be an interesting endeavor.

However, I am a proponent of artificial data, and I firmly believe that the use of artificial projects is optimal for testing new scheduling algorithms. I believe that the true value of empirical projects lies elsewhere in academia, as will be discussed in Chap. 9, and that is precisely why José and I solely relied on artificial projects in the upcoming Chaps. 3 to 8.

2.3 Just a Beginning

The purpose of the two summary articles mentioned at the beginning of this chapter was to provide academic researchers with a clear understanding of the available project data (both artificial and empirical) and give them the values of the network and resource indicators for each project. We thought that compiling everything into two summary publications would eliminate any doubt about where to find the projects and how to use them in academic studies.

However, the artificial project data summary quickly seemed to become redundant because before we knew it, the article was already outdated. José and I soon identified numerous new research ideas that we could test on the current datasets, and with each new idea came the need to further expand the existing artificial project datasets. Therefore, the following chapters present the expansions and adjustments to the existing artificial data, and we no longer recommend readers to refer to the 2016 overview article. Instead, it is much more interesting to turn the page of this book and read further about the quest for high-quality project data.

The progress with the empirical projects was also swift. The empirical dataset from the 2015 article was systematically expanded to eventually include 181 projects which I summarized in the book "*The Illusion of Control*." At the time of writing this current book, the empirical dataset already contained 199 projects, and I would not be surprised if this number has already changed by the time that you read this book. Unlike the first summary article, I believe that the empirical data article remains highly relevant for academic research, primarily because there are few other real datasets accessible to researchers.

We used to hear regularly that too much project data would hinder focusing on the essence of academic research, but we believe that nothing could be further from the truth. It was precisely because of the overwhelming number of artificial projects that our new research ideas emerged quickly. And with each idea came the need for other, better, and thus new projects. Our quest for high-quality data became an adventure with faster computers and many research visits to Ghent and Lisbon. Above all, it became an infinite drive to constantly question and review our data extensions to eventually realize that the quest will never end.

The result is a wealth of additional project data waiting to come in the next chapters.

3

Do We Have Good Schedules for the Projects?

Note: This chapter is based on the article "*A tool to test and validate algorithms for the resource-constrained project scheduling problem*", published in *Computers and Industrial Engineering*.

With the 10,940 projects from six different artificial databases and the 51 real projects from our empirical database, we were able to get started. The goal was not to create as much project data as possible, but to maximize the diversity of the projects so that all researchers, including José and myself, could develop better and faster algorithms to solve the resource-constrained project scheduling problem as effectively as possible. To assist other researchers as effectively as possible, we needed to make the project data easily accessible and provide a platform that allows the projects to be downloaded in a convenient and familiar format with minimal effort.

But that was clearly not enough.

3.1 Where Is the Data?

We considered it necessary to create a website that we could reference in our studies so that researchers could find their way to our projects. The PSPLIB had such a website available, and it was highly popular among researchers. Therefore, we created a similar webpage to add our own project data, which you can find at www.projectmanagement.ugent.be/research/data. However, José and I wanted to take it a step further and dreamt of a tool to facilitate the testing and validation of new algorithms. This would enable researchers to easily access the current state of known schedules, thereby facilitating their search for new and improved solutions for the RCPSP. The new idea and its accompanying tool were placed on our webpage www.solutionsupdate.ugent.be.

From now on, everyone had access to many projects and their best known schedules. If that is not a way to share results!

The SolutionsUpdate weblink is not an easy-to-use website for quickly downloading some data and immediately getting started. It requires some understanding of its operation before you can benefit from it. Any researcher can upload and download a lot of information there, and it is best done with care. In our study mentioned at the beginning of this chapter, we explained that there are three different types of files available on the site. The so-called *instance files* are the files in the known Patterson format containing the project network and resource data. The "*dataset files*" are summary files with various additional data about the project instances of a certain dataset, such as the values for the network and resource indicators. Finally, there are also "*results files*" available, which describe the best known schedules for each project. This enables future researchers to study them and hopefully improve upon them with their new algorithms to solve the RCPSP.

Sharing the projects with the research community is important, but we believe that the greatest advantage of the website is the availability of solutions for each project. As mentioned earlier, merely providing project data was not the main goal of our research, but rather, we aimed to create a healthy yet competitive environment so that more and more researchers could better test their algorithms. The main goal was, and always will be, to increase knowledge about the RCPSP, and for that, a little competition was necessary. Therefore, we made the known solutions (the schedules) of the current project data available in three ways:

- **Solutions:** For most projects, both lower bound and upper bound values are available. The lower bounds provide an underestimate of the planned project duration, and the critical path length is obviously the most obvious one. But more and more researchers were trying to create better lower bounds for projects, and many of them can be found on the website. Much more important than lower bounds are, of course, the upper bounds, which represent a feasible schedule, along with their corresponding makespans. These upper bound values indicate the best known durations for the projects, and it is these values that must be improved by new algorithms, resulting in better resource-feasible schedules.
- **Details:** In addition to the aggregated best known lower and upper bound values for all projects, it is also possible to break them down into values generated for each individual algorithm used to solve the projects. Usually, researchers are not very interested in these values per algorithm, since the main goal is the improvement of schedules found by any algorithm, rather than focusing on enhancing the schedule found by one particular algorithm. However, despite this, it can sometimes be very useful to refine the schedule produced by one particular algorithm, as comparing solutions between different types of algorithms is not always fair. There are different classes of algorithms to solve the RCPSP, ranging from fast priority rules (providing fast and fairly good solutions) to meta-heuristics (providing slower but much better solutions) and even exact methods (providing very slow but optimal solutions). A researcher proposing new priority rules would obviously have little benefit from a comparison with the best possible solutions (which may have been found by an exact algorithm) and would primarily want to demonstrate that these solutions perform better than solutions found with other similar priority rules. Therefore, solutions per algorithm can be useful too, although we must admit that there currently is not yet much information available because we must rely on the goodwill of the researchers to upload their results obtained

from their specific algorithm. This book is a call to all interested researchers to actually do so.

- **Status:** Last but not least, the website also indicates whether a project instance is still open or closed. A *closed* instance is a project for which the lower bound and the upper bound are the same, and thus, for which it is proven that the available solution is the best possible one. Although solved to optimality, these projects remain interesting for researchers because it is often unknown how this solution was found. Perhaps it was after weeks of calculations on a supercomputer, and it may possibly be a challenge to find that same solution with a new algorithm in a matter of minutes or seconds. But the so-called *open* instances are undoubtedly the most interesting projects because they have a known feasible solution (i.e., an upper bound) but without knowing whether it is the optimal one. The quest for optimal solutions is an eternal pursuit of a researcher, and every time an optimal solution is found, researchers go wild with joy. José and I have made quite a few attempts to close some open instances, as you will read in later chapters, but it is a real challenge, and so we had only moderate success. The open instances are, after decades of research, still open for a reason, and so they often consist of very complex projects.

At the time of the website launch, we strongly believed in its potential to increase the availability of project data and facilitate the sharing of solutions to stimulate further research. Since then, there has been traffic, but not as much as we hoped. A few researchers have uploaded their solutions, but it has mostly been José and me using our own website, with not too many others. We somewhat understand that, of course, as researchers want to develop new algorithms and publish them, and feel little need to go through the whole maze of procedures to upload their solutions to a website. However, we still believe that this website will gradually be used more to upload new and better solutions, and we encourage researchers to take a moment to browse the site to see how far academic research has progressed.

I realize that I may have caused some confusion, as I have mentioned two websites already. Therefore, let me provide a brief clarification. If you are a researcher looking for our project data along with the accompanying spreadsheet containing all the values for the network and resource indicators, then I recommend the following website (I have already mentioned it, but I want to emphasize it again):

www.projectmanagement.ugent.be/research/data

Unlike the SolutionsUpdate website, there is no two-way traffic possible (only downloading, no uploading), but it is probably just a matter of a few minutes before everything is on your computer, and you can get started right away.

Only when you have found new solutions or you want more detail per algorithm and per dataset, is it time to explore the challenging world of optimal solutions, searching for details about the data and algorithms, and much more. Then, and only then, it becomes interesting to visit the www.solutionsupdate.ugent.be website. You may spend half a day there, but we promise you that there is much more available than you ever thought possible.

To fully convince the reader of the importance of having a website displaying solutions of existing projects, it is interesting to mention that we were able to find 16 new lower bounds, 37 new best known solutions (upper bounds), and for four projects, we were also able to prove optimality (thus transitioning these projects from open to closed status). And all of this was achieved by only using existing procedures. It seems that the convenience of having an overview of existing solutions indeed sometimes leads to new and better solutions.

3.2 More Is Better (*Sometimes*)

I am well aware that I have written several times that it was not necessarily our intention to generate more project data, while up to now we have actually done nothing else. What I obviously meant is that it is not so easy to claim that having more projects available always leads to better research. Every new project dataset must have a certain rationale, and without realizing it ourselves, José and I kept getting more and more ideas around the RCPSP that consistently yielded new projects. Perhaps more project data is always better after all. Many of those ideas, along with additional data, are discussed in the following chapters, but I want to describe one of our ideas already here that arose while we were working with two new researchers from China.

These two students, Weikang Guo and Jingyu Luo, joined our research group and expressed their interest in using machine learning techniques to solve the RCPSP. Weikang was particularly interested in selecting the best algorithm to solve the RCPSP and needed a large training set of projects with known solutions obtained from various procedures. Despite having many projects available on our website, many of them were still unsolvable (open instances), and for training, she really needed a lot of projects with known optimal solutions. She pleaded with us for more data. Jingyu's story was quite similar, as he became interested in building new algorithms to solve the RCPSP (rather than automatically selecting existing algorithms) based on machine learning methods. Just like Weikang, he expressed his desire for a lot of projects that could be easily solved to optimality. José and I told them that we already had more than 10,000 projects, but we could not convince them that this would be enough. More data. More projects. That is what they kept telling us. And that is what we gave them.

In a way, their request for additional data came as a godsend. José and I realized very well that with the creation of our new websites, we could never achieve a publication. We were looking for an additional contribution to be able to publish our new study. Thanks to the possibility of creating a new dataset, we had found that contribution, and not much later, the paper was published (as was mentioned at the beginning of the chapter).

We could easily have generated additional projects with our RanGen generator, but this time we decided to approach it completely differently. Instead of generating new projects, we decided to expand the existing projects we had so that they could be used for machine learning training purposes. After all, we already had a dataset available where the network structure was varied very accurately based on the four network indicators SP, AD, LA, and TF. The attentive reader knows that I am referring to the MT dataset from Chap. 2, where the network structure is very diverse, but the resources are simply not available.

3 Do We Have Good Schedules for the Projects?

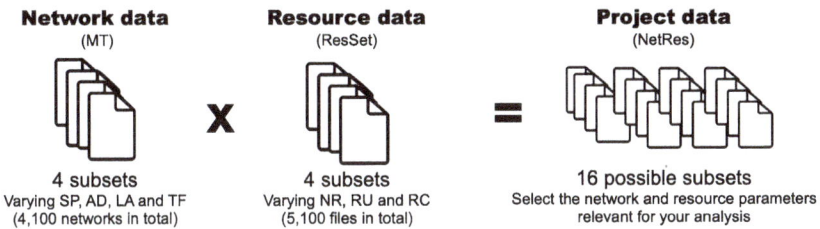

Fig. 3.1 NetRes dataset

	MT set (Network topology)						ResGen set (Resource scarceness)			
	SP	AD	LA	TF	Total		#Res	RU	RC	Total
Set 1	0.1 to 0.9 (Steps of 0.1)	random	random	random	900	Set 1	4	2 or 4	0.25, 0.50 or 0.75	600
Set 2	0.2 or 0.5	0.2 to 0.8 (Steps of 0.2)	random	random	800	Set 2	4	2 or 4	0.1 to 0.9 (Steps of 0.1)	1,800
Set 3	0.2, 0.5 or 0.8	random	0.2 to 0.8 (Steps of 0.2)	random	1,200	Set 3	10	2, 6 or 10	0.25, 0.50 or 0.75	900
Set 4	0.2, 0.5 or 0.8	random	random	0.2 to 0.8 (Steps of 0.2)	1,200	Set 4	10	2, 6 or 10	0.25, 0.50 or 0.75 (Variability between resources)	1,800

Fig. 3.2 NetRes dataset details

We used this set to augment it with detailed data for resources, which we generated in a separate file containing diverse values for the known resource indicators. We varied not only the number of resources (#Res) but also the resource usage (RU) and resource constrainedness (RC) to ensure that all resource possibilities were generated. We named this file containing resource data the ResSet dataset, as it solely contained resource data without project networks with activities and precedence relations. By merging the networks from the MT set with the resource data from the ResSet set, we were now capable of generating a significantly large number of diverse projects with resources. Eventually, we renamed this combined dataset the NetRes dataset (with Net representing networks and Res for resources), as illustrated in Fig. 3.1.

The reason that we had kept the networks and resources separated into two distinct sets was because there were incredibly numerous possible combinations, and the possibilities for project diversity now seemed endless. The MT dataset consists of 4100 networks with 30 activities and comprises four sets with different values for the network indicators as shown in Fig. 3.2. The resource files from the ResSet database contain resource data for exactly 30 activities (because the networks of the MT set have exactly 30 activities) where the number of resources varies between a minimum of 2 to a maximum of 10. The ResSet database is split up into four subsets as well, with varying values for resource use (RU) and resource constrainedness (RC), resulting in 5100 resource files. By combining all these, up to $4100 \times 5100 = 20{,}910{,}000$ possible projects can be generated with a diverse network and resource structure. Obviously, no one needs so many projects, but the possibility is there, and researchers can now assemble their set according to their preference and research topic. At least, Jingyu and Weikang were happy after all.

To avoid intimidating potential users with our large number of new projects in the NetRes dataset, we manually selected 3810 networks ourselves in such a way that the projects contain a maximum diversity in both the network and resource structure, and we placed them in the 1kNetRes set. This increases the number of projects from 10,940 to 14,750 projects (with the option to generate many more projects), and we believed that we now had certainly enough projects.

3.3 No More Data

Indeed, that is what we thought.

Our study of this chapter was completed at the beginning of 2015, although we only decided to write down our results in a paper two years later, in 2017, which we submitted to the *Computers and Industrial Engineering* journal (and which was eventually accepted one year later). By 2015, I had already known José for thirteen years, and we had achieved a lot in that time. We were happy that our first meeting in Valencia in 2002 had led to new project datasets, two new websites, a number of publications, and an intense friendship that would never fade away. But as far as the data research was concerned, our story was more or less told. With the 14,750 available projects, researchers could reignite the search for better solutions for the RCPSP, and we were convinced that there would be no need for additional projects in the coming years. We hoped that our collaboration would not stop here, but we feared for a moment that it would not amount to significant progress for the time being.

No more data. Our story was told.

Thankfully not. The radiant sun from Lisbon had somehow seduced me into never forgetting the city, and I had learned that rainy weather (as is often the case in Belgium) was not necessary to spend hours behind the computer for rigorous research. Perhaps it was time for a period with only research and no teaching. At the faculty where I work (in Belgium), they introduced the option to disappear for a while to allow for quiet contemplation of further research, but doing nothing—except thinking about future research ideas—seemed like pure horror to me. I asked the then-Dean if the definition of a sabbatical could also be interpreted slightly differently, as a period to increase the intensity of academic research. Instead of entering a relaxed thinking mode, I enthusiastically told him I wanted to seek a solution to one of the most challenging problems in the literature. The Dean had never heard about the RCPSP but said I should do what I could not resist, provided it produced output, and he gave me a ticket to stay in Lisbon for 2 years.

I promised him that I would create a super-fast algorithm to solve this challenging project scheduling problem better and faster than ever before. Where could I better carry out this new challenge than in the most beautiful city in the world, where someone lives with whom I had the most intense and joyful collaboration in the past thirteen years?

In 2015, I decided to move to the city of light for two years.

It became a true and intense research adventure, one that I can still enjoy to this day.

4

Can We Solve Every Project Instance?

Note: This chapter is based on the article "*An exact composite lower bound strategy for the resource-constrained project scheduling problem*", published in *Computers and Operations Research*.

In the previous chapters, the focus was primarily on our quest for new and better project data. However, it is in this chapter that the collaboration between José and me truly took off. We were no longer in need of additional project data, and we had decided to use all of our available projects to solve the RCPSP once and for all.

Since I was on sabbatical in Lisbon, we met frequently, at least once a week, in an old but beautiful building close to Jardim do Príncipe Real. We discussed all kinds of new ideas that went in all possible directions, but ultimately, it was a second life-changing event that set us on the right path.

4.1 Life-Changing Event #2

We decided to return to our old first shared research passion and had the ambition to solve all projects to optimality.[1] Both of us had worked on branch-and-bound methods to solve the RCPSP when we were PhD students, and we noticed that these algorithms were not as widely used in 2015 as they were when we were young doctoral students. The rise and popularity of meta-heuristics might have had something to do with this, and more and more researchers were diving into that area of research. In a certain way, it was a logical evolution. Computers were becoming increasingly powerful, and meta-heuristics performed quite well under this increased computing capability. Moreover, the pressure to publish papers had become greater than ever, and since meta-heuristics provided solutions that were close to optimality, there were few incentives left to embark on the difficult quest for a better branch-and-bound method and search for the optimal solution.

However, when we were PhD students, times were different.

In the late 1990s and early 2000s, there were quite a few researchers proposing new branch-and-bound methods in their studies, and we were two of them. At that time, one of the most well-known branch-and-bound methods was the one proposed by Erik Demeulemeester in the flagship journal *Management Science* in 1992. The method is, until today, considered as one of the best performing methods to achieve optimal solutions for the RCPSP, at least for small projects.[2] But of course, there were several other studies published with branch-and-bound procedures to solve the RCPSP, each quite powerful, but only slightly different from the others, often in the details. José and I had embarked on a plan to program and test all these methods on our new project data. Most of these methods were developed more than a decade earlier, and with our new and faster computers in 2015 and much more project data, we were convinced

---

[1] That turned out to be a bit too ambitious, in hindsight, but sometimes you just have to set the bar high and be a little naive in academic research.

[2] The attentive might have noticed that this is the procedure we also used in Chap. 1 for finding phase transitions in the project data.

that we would be able to solve almost every project to optimality. Our ambition was endlessly vast, even a bit naive maybe.

But there was no lack of enthusiasm. When I began my sabbatical in 2015, I had hoped that I could occasionally enjoy the sun in Lisbon, but the reality was that with that new idea, José and I rarely emerged from our dark research cellar. Solving every project optimally turned out to be not as straightforward as we initially thought. We had indeed computers that were much faster with a RAM memory that we could only dream of during our doctorate, yet various difficulties arose. At the time when most branch-and-bound procedures had been developed, tests were mainly conducted on the Patterson set, which consisted of 110 instances. A few years later, researchers also tested these methods using the J30 set from PSPLIB, which comprised 480 projects, and that was essentially the extent of it. If we could reprogram all of these procedures, then we could also test them better this time. However, we had so many more projects at our disposal, most of which were not much larger than 30-activity projects, and with so much data, it became quite a challenge to test everything. It seemed like an impossible task, despite the fast computers we both had.

And then I told José about the supercomputer at Ghent University, and our research life changed for a second time.

I have rarely seen anyone as enthusiastic as José when it comes to discussing computer details. I had conducted a few experiments on this powerful computer myself, and it seemed useful to do more with it, but José started asking me all sorts of questions about this powerful machine to which I usually could not give a reasonable answer. He was excited about the number of cores available, about the way this computer could be operated remotely, and about how they managed to keep its heating problems under control. "Think about all the computer experiments we can run with this computer," he enthusiastically said. "It's almost unbelievably endless!" I was certainly excited too, but for José, it was like a whole new world opening up. One that was going to change everything.

And then we took the most important decision in our research career, and I appointed José as a guest professor at our faculty in Belgium which included remote access to this supercomputer infrastructure. Looking back on this decision, we can confidently say that it was one of the most rewarding decisions of our careers. From then on, everything indeed changed. José explored the dark secrets of parallel computing and scripting and continues to enthusiastically share all the details with me to this day. Then I mostly nod, smile, and pretend to understand but the truth is that José is the one of us who manages these dark secrets of computer power. This supercomputer has rarely been idle since that day. It was from that decision on that I realized why the chemistry between us is so magical. We are united by a fundamental shared passion for project scheduling, fuelled by our collective expertise, knowledge, and deep insights into the field. While our enthusiasm knows no bounds, it is in the nuances where José's strengths complement mine, and it works in both ways and adds the finishing touches to our collaborative efforts.

I guess that if you meet someone in your line of work with whom you are a perfect fit, you are very fortunate. That magic happened to us in the most beautiful city in the south of Europe.

4.2 From 12 to 48 Algorithms

That is how our ambitious adventure began with the challenging task of thoroughly studying each existing branch-and-bound procedure in detail and subsequently programming them ourselves. We reviewed all studies in the literature, as is customary for academics, and found that a total of 12 known branch-and-bound procedures had been published in the past. After a thorough study, we were able to map out the similarities and differences and identify the five main components. These components are present in every branch-and-bound procedure, albeit implemented in different ways, and are briefly explained below. As we began to understand better how these components were assembled, we became convinced that they could be combined in ways never done before, allowing us to create more versions of the branch-and-bound methods beyond these 12 existing ones. If you are a newcomer to the world of branch-and-bound methods, then the description of the following five components can be used as a sort of crash course tutorial to get acquainted with this powerful technique:

- **Tree strategy:** Since a branch-and-bound procedure divides the problem to be solved into smaller subproblems (branches), this search is represented in a tree. The strategy to explore the tree can be done using a "depth-first" strategy, which explores one path in the tree as deeply as possible before backtracking and trying another path. The counterpart of this method is known as "breadth-first" where all possible paths (or branches) at the same depth level of the tree are explored before moving deeper into the tree. A hybrid form of these two extreme methods is the "best-first" strategy, which selects the most promising (best) branch each time, regardless of the depth of the branch in the tree. There are also hybrid forms of these methods available, but since it has been well established in the literature that the depth-first strategy performs the best, we decided to limit ourselves to this strategy and leave the others aside.
- **Search strategy:** This depth-first strategy can be implemented in different ways. The so-called upper bound strategy (U) searches for feasible schedules which act as upper bounds on the project makespan, and the search continues until the lowest upper bound value is found, representing the best possible feasible schedule and therefore the optimal solution. This strategy is most commonly used in the literature. However, there is also a promising alternative strategy called the lower bound strategy (L). With this strategy, it is assumed that a certain feasible schedule exists with a duration of X time units, which acts as a lower bound value on the project makespan. The branch-and-bound procedure then tries to find the schedule with a makespan of X time units during its search. If no solution can be found, it simply means that no solution with a duration of X exists, and the method starts the search again assuming a schedule exists with a duration of X + 1. This process is repeated k times until a feasible schedule is found with a makespan of X + k time units, which then guarantees the optimal solution (since no other schedule exists with a lower makespan).
- **Branching scheme:** I have already mentioned that a branch-and-bound procedure generates several nodes at each level of the search tree by breaking down the problem into subproblems. While the previously discussed tree strategy determines the construction of the

tree by deciding the order in which nodes are explored, the branching scheme however determines how each created node will be divided into child nodes. The simplest but not widely used approach to split nodes involves branching on the activity start time (A), where a specific activity is chosen to branch, and then child nodes are created for each possible start time of this activity. An alternative and more effective approach is the serial branching scheme (S) which iteratively selects activities that are eligible for scheduling using an activity incrementation scheme. More specifically, this method looks at the activities that are already scheduled at the current node of the tree and then selects all activities for which all their predecessor activities are already in this partial schedule. Each of these activities can then be scheduled as early as possible in the partial schedule without violating the resource availabilities, which creates several child nodes in the search tree. However, the most commonly used method is the parallel branching scheme (P), which uses a time incrementation approach by iteratively increasing the time in the schedule and searching for eligible activities that can be scheduled at that time point without exceeding the resource availability. Since this can be done under various combinations of activities, it creates different child nodes for further branching.

- **Branching order:** When the branching scheme has split a certain node into a number of child nodes, the algorithm must make a decision about which child node is chosen first for further splitting (and which one thereafter). Therefore, the child nodes are ranked based on a criterion so that this decision can be made automatically. There are many possibilities, but the most common method is to make this decision on the lower bound value of the nodes (B) which provides an estimate for the smallest possible makespan of the schedule that can be created from this node in the tree. The best choice is then to explore the node with the lowest lower bound value first. Because there are many other options for making this choice, we also included a ranking based on minimum time windows of activities (M), but since it was only used by one existing algorithm, we chose not to implement it. Our experiments were able to demonstrate that the branching order does not always have a significant effect. Therefore, we also included a random choice by using the activity number in the project network (A) to add some diversity to our algorithm.

- **Lower bounds:** The use of lower bounds to select child nodes is predicated on estimating the minimum duration of the project for each subproblem represented by the nodes in the search tree. While such lower bounds are, of course, mere approximations and entail a computational overhead, it is not unreasonable to prioritize the child nodes with the lowest predicted duration as proposed by the lower bound branching order strategy. After all, that node is more likely to yield a favorable schedule than a node with a significantly longer minimum duration prediction. The lower bounds can also be used to determine whether it is useful to further investigate this node in the branch-and-bound tree. When the lower bound value is equal to or greater than the makespan of a previously found schedule of the search, this node should not be further evaluated, as no solution better than the already found solution can be obtained from this node. In this way, many parts of the tree can be pruned, resulting in a faster search for an optimal schedule for the project. Given the time-consuming nature of lower bound calculations for each node in the tree, many branch-and-bound methods typically rely on the easy and fast critical path lower bound, or its extension to the somewhat more intricate critical sequence lower bound, but not beyond. However, the literature offers superior albeit slower lower bound calculations, prompting our decision to incorporate no less than 13 lower bounds in our branch-and-bound method, categorized into four classes. We always used the critical path bound (indicated by bound 0), but we also added 4, 8, or 12 other lower bounds in our algorithm, hoping it would make our branching order choice much more effective.

Each existing branch-and-bound algorithm had implemented only one possible combination of these five components that apparently led to good results, but logically, there were many more combinations possible. Ultimately, we decided to combine all possible components in the following way (using the abbreviations used in the description in the previous text section):

$$[U,L] \times [A,P,S] \times [B,A] \times [0, 4, 8, 12].$$

This led to our 2 × 3 × 2 × 4 = 48 possible combinations, which are actually 48 different types of branch-and-bound procedures, although they are sometimes quite similar and only differ in the details. But we learned many years ago that details can make a big difference in academic research.

With these 48 methods, we had almost all 12 existing branch-and-bound methods from the literature at our disposal. Each of these 12 procedures only incorporates one possible combination of the 48 possibilities, while our own implementation with the 48 possible combinations is much broader than the entirety of all existing procedures. However, we must mention that for two of the twelve procedures, there were other special components present that we have not implemented. This means that we still had 10 out of the 12 procedures included in our 48 branch-and-bound procedures, providing some assurance that our experiments could almost perfectly replicate the search for optimal solutions as if all current state algorithms of the literature were used.

Moreover, our implementation is the only one to use up to a maximum of 13 lower bounds for exploring the branch-and-bound tree. However, such a decision naturally has repercussions on the computational time, thus requiring some additional explanation. The challenge of using lower bounds in a branch-and-bound procedure is that a trade-off must be made between the quality of the lower bound and the computation time required to calculate its value at each node of the tree. While all procedures from the literature only included one or at most two lower bounds in the method, we decided to use all lower bounds proposed in a study by Klein and Scholl in 1999. It is quite easy to understand that this choice can obviously improve the quality of the lower bound estimates, but the computation time can skyrocket at times, making it impossible to implement this for each node of the tree.

That is why we decided to incorporate an automatic trade-off analysis between quality and speed in the branch-and-bound method by introducing a so-called credit/tabu list system. It is not very complex, but it definitely works. Initially, at the start of the search, all 13 lower bounds are calculated at each created node to estimate the minimum project duration, which takes quite some computation time. For each calculation, the quality of each lower bound is monitored and saved.

Lower bounds showing higher values are the better lower bounds, but those that consistently yield poor values are set tabu for a while, meaning that they will not be used for the next nodes to save some computation time. After the tabu time expires, they are reintroduced in the lower bound calculations in the hope that they will then yield better values for other nodes in the tree (but if not, they are put back on the tabu list for a while). This way, computation time is not wasted on unnecessary calculations, making the branch-and-bound procedure more efficient. The credit list operates on the same principle but in reverse. If a certain lower bound yields a very good value for a node, it is placed on the credit list for some time so that it will always be used for the next created nodes, even if it performs not so well there. In other words, the credited lower bounds can never be put on the tabu list as long as their credit values do not expire. As explained earlier, the choice of which lower bounds to incorporate into the tabu/credit system always included the critical path lower bound, and possibly extended with 4, 8, or 12 additional lower bounds. By designing such an automatic trade-off between lower bound quality and corresponding computation time, we decided to rename our procedure into the branch-and-bound procedure with a composite lower bound strategy.

Programming all these combinations took a while, but one of José's strengths is that once he starts programming, he simply does not stop. With this new extended procedure programmed in C++, it was time to harness the full glory and power of the supercomputer and get to results. We wanted to solve all projects optimally, so exciting times lay ahead of us. It was summer and hot in Lisbon, and we could certainly use some new research breakthroughs to cool down a bit.

4.3 Solving Instances to Optimality

What a summer it was. I cannot quite remember how many experiments we conducted, but there was an enormous amount of them. We did not only use the PSPLIB dataset for our test experiments, but also used the RG30, RG300 and even a portion of the 1kNetRes projects. It is no surprise that with such a large number of projects, each experiment soon consumed more than a year of computing time (if conducted on our laptops). I understood now why José was so enthusiastic about that powerful Ghent supercomputer, as tasks that could take days could be completed in just a few minutes. It is incredible what such powerful machines can do for academic research, and I cannot remember how we managed to carry out experiments during our doctorate with computers that had hardly any memory and no processing power. Perhaps I sound a bit like an old man now, waxing nostalgic about the past.

I am sure you empathize with the enthusiasm we felt back then, but the results of our experiments were initially a bit disappointing, as we quickly realized that we would not be able to solve all open project instances to optimality. In hindsight, it may have been somewhat naive to think that all projects would be optimally solved only because we had a powerful supercomputer, and our ambition soon gave way to realism. Eventually, our disappointment was turned into joy, as we could find 135 new lower bounds and 5 new optimal solutions for the PSPLIB set, after many other researchers had been searching for such solutions for years but never found them. For the other sets (RG30, RG300, and 1kNetRes), we could find many optimal solutions, but that was less impressive as not many researchers had attempted already to solve these project instances. But still, not a bad outcome in the end.

We also investigated why certain procedures could solve some project instances, while others could not, and what particularly surprised us was that many of the projects with 30 activities could not be optimally solved at all. The reason for our surprise was that all J30 instances from the PSPLIB set had already been optimally solved many years before our research. Many researchers seeking optimal solutions for the RCPSP therefore turned to the J60 instances from the PSPLIB set, which were larger and thus more difficult to solve. Gradually, researchers began to believe that the procedures had become increasingly powerful to assume that all 30-activity projects could be easily solved to optimality. But that was clearly not the case. For many of the RG30 instances, especially those with a highly parallel network structure, optimal solutions could not be found even after hours of searching, not even with 48 different implementations of a powerful technique.

4 Can We Solve Every Project Instance?

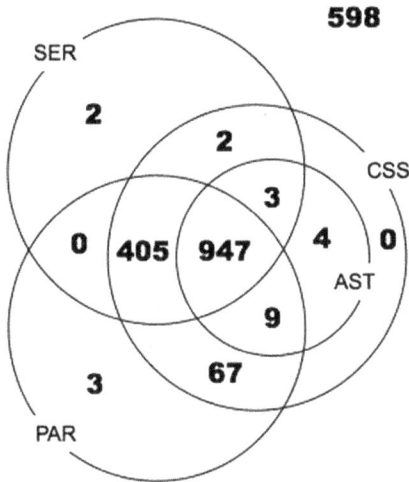

Fig. 4.1 Number of solved PSPLIB instances

When we examined the contribution of the five components more closely, we were delighted to see that no procedure from the literature could dominate the others. We were delighted because this meant each procedure in the literature had a unique contribution that others did not possess. Apparently, only through collaborative research studies by various academics, each proposing slightly different approaches to solve the RCPSP, can the RCPSP be addressed more effectively. A compelling demonstration of the potential to combine existing procedures and extend them, as we did, is depicted in Fig. 4.1, which displays the number of solved instances from the PSPLIB set truncated after 1 hour of computer time per instance. The figure shows the three used branching schemes—the serial, the parallel, and the activity start time branching—and our new combined search strategy (CSS). The CSS strategy is quite simple, dividing the total 1-hour search time into three equal intervals of 20 minutes each. It varies the branching scheme from the serial (SER) to the parallel scheme (PAR), and eventually to the activity starting time (AST) strategy.

The figure shows that 598 instances could not be solved to optimality by our experiments, regardless of the chosen method, but only 5 solved instances could not be solved by the CSS strategy.[3] Moreover, the figure shows that each individual strategy performs much worse than the combined strategy. 83 solved instances could not be solved by the SER strategy, 11 by the PAR strategy, and 479 instances by the AST strategy, and that is much worse than the five instances for the CSS method. These results demonstrate that a thoughtfully crafted combination of existing solution procedures holds promise for uncovering better solutions to a complex scheduling problem like the RCPSP. Ultimately, the primary goal of academic research is indeed to discover significant improvements through various research studies, and in this case, that goal has certainly been achieved.

Many years later, when Weikang joined our research team,[4] we decided to fully exploit this potential by automatically selecting the best performing procedure based on network and resource indicator values for each project. More specifically, we relied on various machine learning algorithms to train our project data, hoping that the values of the network and resource indicators for each project could be used to select the best of the 48 branch-and-bound methods. Results of new experiments showed that these machine learning methods always outperform any single best branch-and-bound method and often come close to the best possible solution that could be obtained by running each of the 48 methods separately at the cost of much higher computation time. The study was conducted many years after the study of this chapter, but it once again illustrated that combining solution procedures in a well-thought-out way is likely to lead to better results.

4.4 Hunger for More

Time flies when doing research, and it was already the second year of my sabbatical. The year of 2016 came to an end when we finally had the results

---

[3] The parallel branching scheme was able to find optimal solutions for three projects that could not be found by the CSS strategy, and the serial branching scheme found two of them. However, the AST strategy was completely dominated by the CSS strategy.
[4] I introduced Weikang already in Chap. 3 as you might remember.

fully written to our satisfaction in a paper. We decided to submit it to the *Computers and Operations Research* journal, which, thanks to the editor's critical yet efficient approach, was known as a very reputable journal. The revision process at that journal also took some time. But after several intriguing adjustments suggested by the reviewers, our study was finally published in 2018, and we were immediately excited. The reviewers' comments were not only challenging but also highly educational, and we were determined to submit more of our research studies to this outstanding journal, as you will see in the upcoming chapters.

Revisions in journals can be frustrating at times (well, most of the time, actually), but not for this submission. We learned a lot from the reviewers' comments, and although we were pleased that they eventually decided to accept our paper, they left us with many new questions. We were still puzzled by the disappointing results for the 30-activity projects and still had not grasped why some projects were easily solvable while others were not, even though they sometimes did not differ much from each other. It was evident that the availability of numerous algorithms and powerful computers was not sufficient to optimally solve all small projects, and that ignorance continued to haunt us.

Apparently, small projects are also often difficult to solve, and we were eager to understand why.

Maybe it was due to the network structure, as parallel projects are typically harder to schedule than serial ones. Alternatively, it could be attributed to the limitation of resources. When resources are medium to very scarce, the problem is almost unsolvable. Perhaps there was indeed a clear yet unknown reason, but truth be told, we did not really have a very good idea of what it could be.

Before returning to Belgium to resume my normal duties, we agreed that José would take a summer sabbatical in Belgium the next year to seek answers to these questions. I warned José that our summers are not as sunny as those in beautiful Lisbon, but that did not deter him. Perhaps he was just excited at the prospect of seeing the supercomputer in action.

Less than one year later after the submission of our work, in the summer of 2017, José was in Ghent, and we were back together in our weekly meetings with a new quest and the supercomputer at hand. Because we wanted to understand why small projects were sometimes difficult to solve, we decided to search for many more small yet challenging projects. And so it was time to generate new project data once again.

Perhaps we should have known that our 14,750 projects from the previous chapters were simply not enough to understand the complexity of the resource-constrained project scheduling problem.

It seems that this challenging scheduling problem is not inclined to easily reveal all of its secrets.

5

Why Is the RCPSP So Difficult? (Part 1)

Note: This chapter is based on the article "*Going to the core of hard resource-constrained project scheduling instances*", published in *Computers and Operations Research*.

During José's sabbatical in Ghent, we developed numerous ideas to understand why the RCPSP is so difficult to solve for some of our projects, but most of these ideas yielded little progress. When he returned to Lisbon at the end of the summer of 2017, we had not many results, but at least, we had compiled a long list of possible research topics to work on in the following summers. We made it a regular habit to visit each other during the holidays, and so we spent a few weeks of research time in each other's cities during the summers of 2018 and 2019. Many of our ideas for this study also originated from our discussions at the 16th Conference on Project Management and Scheduling in Rome in 2018. It is my personal opinion that academic conferences are not only interesting for seeing the work of other researchers (which is important, of course), but also because you are immersed in a research atmosphere for a few days, where often the best ideas emerge. José and I talked a lot about our research during this conference, with the Colosseum and the Trevi Fountain in the background. The result of this collaboration is described in this chapter.

From our studies of the previous chapters, it had become clear to us that it is not easy to understand what makes a project difficult or easy for the RCPSP. Nevertheless, there must be some characteristics, which can possibly be measured by the previously discussed network or resource indicators, that are responsible for the complexity. If we knew those, it would be easier to predict a project's complexity, and then we could align our algorithms much better with the specific characteristics of the projects. But clearly, we were not there yet.

We had done everything in our power to solve this puzzle. Even with our 48 algorithms from the previous chapter and numerous experiments on our project data, we just could not explain why some 30-activity projects remained unsolvable, while others with 120 activities were solved in a fraction of a minute. We remained in the dark about the specific elements responsible for the complexity, and so we decided to search for the essence of each project, for that little part of the project that prevents the algorithm from finding the optimal schedule.

We called our idea a quest for the core of the problem complexity.

5.1 Small Projects Are Easy

There was, of course, already a known answer to the question of why the RCPSP is so difficult to solve, but it is not a quite straightforward one to grasp. The scheduling problem is recognized as an NP-hard problem, referring to a class of problems for which no polynomial-time algorithm exists to solve all instances of the problem. Typically, for these problems, small instances with only a few activities can often be solved with simple algorithms, or by brute force, where all schedules are simply generated and the best one is selected. However, as the size of the problem increases, the number of possible solutions typically grows exponentially, making it impractical to explore all possibilities, thus rendering the problem very difficult. For this reason, it is not difficult to understand that the RCPSP may be relatively easy to solve for small projects, but it can quickly become intractable as the number of activities increases. This NP-hard property makes the RCPSP challenging and attractive to researchers, often motivating their choice to develop meta-heuristic methods to find near-optimal solutions within a reasonable amount of time instead of exact algorithms that search for the optimal schedule.

But we were not interested in near-optimal schedules. We were determined to find optimal solutions, even though it proved challenging even for small projects. However, we were determined to pursue this quest for optimality because we had our 48 exact branch-and-bound procedures at our disposal, which, given sufficient computing time, always guarantee optimality. As if the 48 procedures were not enough to embark on this quest, we also programmed several well-known mathematical integer programming models that can also guarantee optimality. All these procedures use lower bound and upper bound calculations for the project makespan, and when these two values for a project are equal to each other, it has been demonstrated that the schedule with the upper bound value is the optimal schedule. That is where we wanted to go!

The objective was clear, but it remained a challenge to clearly delineate when a project instance was difficult or easy to solve. The NP-hard definition from the theoretical computer science discipline provided little insight due to its mathematical abstraction, so we decided to adopt a very simple and pragmatic definition to describe the complexity of a project.

> **Definition**: A project instance is assumed to be hard if it can not be solved to optimality within reasonable time using a state-of-the-art exact algorithm.

With our wide range of solution methods, we assumed that we had state-of-the-art exact algorithms at our disposal. The word *"exact"* is used here to express that the algorithm, given enough computation time, can guarantee optimality. This is the case, for example, for branch-and-bound procedures that we used in this study but cannot be guaranteed when meta-heuristics are used. Moreover, with the supercomputer, we could allocate enough time so that *"reasonable time"* seemed like a relative concept. It was clear that we had sufficient resources at our disposal to experimentally search for optimal solutions to determine whether or not they could be found, thus determining what makes a project complex or easy.

But we were not just searching for optimal solutions for the projects in our datasets, as we had done in our previous study (with moderate success). Our plan was to manipulate and modify the existing projects in a way that would make them both smaller and more difficult. Making a project smaller is an easy task and can be done by removing activities or resources from the project, but usually, these projects then become easier to schedule optimally. Making projects more difficult is also not very challenging, but it usually means that we have to make them larger, and that is not where we wanted to go either. It should be clear that our quest for small yet complex projects was anything but straightforward, and that is why we were so grateful to have both our state-of-the-art procedures and the powerful supercomputer at our disposal.

And so it happened that we once again spent several beautiful summers behind our computers, searching for the secret of the complexity of the RCPSP.

We were in search of the smallest possible projects for which no optimal schedule could be found.

5.2 Going to the Core

We used 8370 projects from our datasets[1] and constantly modified them by making various changes to reduce their size. After each modification, we used our algorithms to attempt to solve the modified project optimally. If successful, we concluded that the modified project was too easy, and we threw it away. If unsuccessful, however, we made further changes to reduce the project's size even more and then again experimentally determined whether the modified project was difficult or easy to solve. We continued this process until we obtained a project containing fewer than 30 activities that still could not be optimally solved by our algorithms. Only then did we classify this project as small and difficult.

---

[1] The projects were taken from the RG30, 1kNetRes, PSPLIB, and DC2 sets.

Since our quest essentially amounts to checking whether the project complexity increased with each modification, it can be very simply summarized in the following pseudocode:

Continue changing if complexity ↑.
Undo the change if complexity ↓.

It was a very simple idea, but it required a lot of computation time and even made the supercomputer sweat.

To structure the number of possible project changes, we implemented five potential modifications for each project. Most modifications typically resulted in a reduction in the project's complexity, as they were all designed to make the project either smaller or just slightly different. However, occasionally, an increase in project complexity was observed, which was what we were looking for. When this occurred, we made further modifications and continued until we had ultimately removed all the easy parts of the original project, leaving only the essence of complexity. It is for this reason that we named our study "*going to the core.*" The five potential modifications are summarized briefly below:

- **Step 1. Remove activities:** In our pursuit of small projects, we aimed to eliminate as many activities as possible without diminishing project complexity. We continued searching for modifications that increased complexity, stopping only when the resulting project had a maximum of 30 activities.
- **Step 2. Remove resources:** Similar to activities, we explored the removal of as many resources as possible to reduce project size. These modifications also typically resulted in simplifications, but occasionally they also led to increased complexity.
- **Step 3. Change resource availability:** The availability of project resources also influences project complexity, as an increasing number of resource conflicts can arise during the search. Therefore, we also varied resource availability randomly in numerous ways, hoping to find more complex projects.
- **Step 4. Change activity duration:** Activity durations can impact complexity too, so we adjusted them both downward and upward, aiming again to identify projects of greater complexity.

	Algorithm	Solution	Time
Phase 1	12 priority rules and 13 lower bounds		1 second to 1 minute
	Composite branch-and-bound		
Phase 2	Composite branch-and-bound	If LB = UB: Stop Otherwise: Phase 3	1 hour
Phase 3	Extended composite branch-and-bound MIP 3 meta-heuristics	If LB = UB: Stop Otherwise: Keep	20 hours

Fig. 5.1 Three phases to search for complex projects

- **Step 5. Change resource demand:** Finally, we manipulated the resource demand for activities to provoke more resource conflicts, thereby enhancing complexity once more.

Since these five modifications were iteratively applied to each project, and since most of them led to simplification rather than the rare instances of complication, it seemed like we were shooting in the dark, hoping to hit a random bird. With so many possibilities and so little chance for complexity increases, we had to strive endlessly to achieve even a small amount of success. Therefore, we decided to consolidate these five modifications into a three-phased procedure, visually summarized in Fig. 5.1 and explained below.

Phase 1. Quick Search Since we had to assess whether the complexity had decreased or increased after every possible modification to a project, we could not always rely on our algorithms due to the significant computation time this would require. Therefore, in this first phase, we made an estimation of the complexity of the modified project using a combination of quick procedures. Specifically, we calculated the lower bound and upper bound of each modified project without spending too much computation time. When these two values are equal, the project is optimally solved and no longer usable for further analysis. However, if these values are not equal, we retained the project for further analysis by making new modifications in Phase 2. We calculated the lower bounds using the 13 lower bounds from the previous chapter, while the upper bounds were computed using 12 very fast priority rules and the composite branch-and-bound procedure (CSS) from the previous chapter, but terminated after a maximum of 1 minute.

Phase 1 is a quick and easy way to filter out many easy projects, keeping only the potentially difficult ones for the next phase. In the next section, I will show that even this quick method was not fast enough, and so we were forced to make further compromises to keep the computation time under control. Despite this, for the remaining projects, there was hope that they could be complex, and this hope was further tested in Phase 2.

Phase 2. Intense Search To determine whether the remaining projects were indeed complex, we needed to compute better values for the upper bounds. Therefore, we again used the composite branch-and-bound procedure, but this time allowed it to compute for a maximum of 1 hour per project (and this for each modification). Our procedure was still making many modifications to the remaining projects, and so this second phase also consumed a lot of computation time. However, at least a significant number of projects had already been removed from our database in the previous phase, justifying the more intensive search. Eventually, the easy projects were removed again, leaving 1315 modified projects available for further analysis in the third and final phase.

Phase 3. Crazy Search In this final phase, we pushed our algorithms to their limits. The 1315 remaining projects were subjected to a very intense search in this last phase by allowing the branch-and-bound procedure a maximum of 20 hours per project. We did not make any additional modifications anymore, as all projects were reduced to a maximum of 30 activities, and we only wanted to test whether these projects were actually complex or not. We used all 48 configurations of the branch-and-bound of the previous chapter, we also employed a mixed integer programming model, and we even relied on three fast and powerful meta-heuristics (which we will further use in Chap. 7). If even with this exhaustive search, no optimal solution could be found for these projects, then we assumed that it would be almost impossible to find one. This phase took 183 months of computer time, but eventually, we were able to retain 623 projects and classify them as *very difficult to solve*. These projects were indeed not only small but also difficult to solve, which ultimately was the goal of this intense search.

5.3 44 Years on a Computer

The entire three-phased procedure cost us more than 40 years' worth of computer power (if we had executed it on a standalone computer). Phases 2 and 3 were very intensive phases and naturally incurred a significant amount of computer time (up to 183 months for Phase 3, for example) but fortunately were only executed on a subset of the projects resulting from Phase 1. The greatest challenge in controlling the computer time lay in the first phase, where so many modifications were performed that even 1 minute per modification quickly resulted in excessively long computing times. The 13 lower bounds and 12 upper bounds for the priority rules could be calculated in no time, but we had to avoid using the CSS procedure for each modification, as it required a full minute, which was far too much.

Therefore, we sought a type of "complexity predictor" that could estimate how the complexity of the project would change with each modification. Only if it predicted an increase in complexity, we proceeded to test it by investing a valuable minute in Phase 1's CSS procedure. Otherwise, we canceled the modified project and proceeded with other modifications.

This complexity indicator[2] consists of a seven-digit number, comprised of three parts, as shown in Fig. 5.2. The lower the number, the easier the project was estimated to be. With each project modification, the digits were filled in to determine the value of the complexity predictor, following the next three-step process:

- **Digits 6 and 7 (CPU time):** The final two digits of the complexity predictor represent the computational time required to determine the

---

[2] In the published version of our study, we referred to this complexity predictor as a *hardness indicator*. However, since it does not directly measure the true hardness of a project but rather makes an estimation of it, I prefer to call it a predictor rather than an indicator. Maybe it does not matter much how it is called, but I find myself thinking deeply about trivial matters sometimes.

5 Why Is the RCPSP So Difficult? (Part 1)

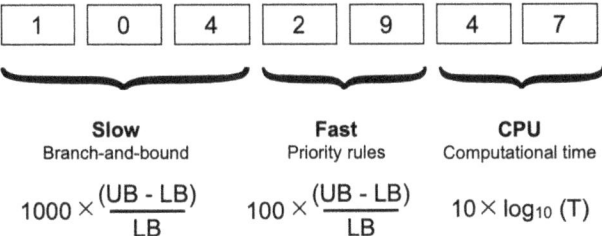

Fig. 5.2 Predicting complexity on the fly

values of the other digits 1 to 5. This time is depicted on a logarithmic scale calculated as $10 \times \log10(T)$, where T is the time expressed in milliseconds. For instance, imposing a maximum time limit of one minute can yield a maximum value of $10 \times \log10(3600) = 47$ for the last two digits. Naturally, this CPU time is only known once the values for digits 1 to 5 have been calculated, but it accumulates gradually and is updated whenever more information becomes available.

- **Digits 4 and 5 (fast solutions):** The value of the solution obtained through the fast priority rules (upper bounds) is compared with the best of the 13 lower bounds and expressed as a percentage difference between the two, multiplied by 100. Since both lower and upper bounds are computed very quickly, this difference can be calculated easily without consuming significant computer time, although it may not fully capture the true project complexity. As this percentage is inserted into digits 4 and 5 of the complexity predictor, it carries a higher weight than the CPU time of digits 6 and 7 but a lower weight than the solutions of digits 1 to 3.
- **Digits 1 to 3 (slow solutions):** The composite branch-and-bound procedure, which is allowed a maximum of 1 minute of computation time, can provide relatively good lower and upper bound values on the project makespan. The difference between these two values, expressed as a relative percentage multiplied by a factor of 1000, is positioned in digits 1–3 of the complexity predictor. It receives a higher weight than the percentage deviation of the fast search (digits 4 and 5). This

is justified because it is obtained through a more intensive search and is therefore more reliable.

This 7-digit complexity predictor may consist of three parts, but for a computer, it is just one number. For example, when the computer encounters the value 1,042,947 (which looks like 104/29/47 to us), it indicates that the total time for finding the best possible upper bound was 60 seconds (10 × log10(60,000) = 47.78, rounded to 47), with a percentage difference of 29% obtained by the priority rules and 10.4% found by the branch-and-bound search. Therefore, the computer evaluates the complexity of a project instance after each modification by simply comparing this single 7-digit number with the value before the modification to determine whether the complexity increased or not. As a result, we can save a considerable amount of time.

I am well aware that I am delving into more detail than usual in this book, but I simply want to emphasize that our quest for small challenging instances would never have been possible without this complexity predictor. It was already difficult enough to identify an increase in complexity after our modifications, so we had to test many projects with numerous alterations. The complexity predictor ensured that the 44 years of computer time did not become tenfold (or more). Therefore, I will briefly outline below an example of how the complexity indicator was used to compare the complexity of an instance 1 with three modified instances 2, 3, and 4. The explanation is almost literally copied from our published paper, something I rarely do in this book, and Fig. 5.3 can also be found there. The reader who is not interested in delving into the details of this predictor can safely skip to the next section without losing the overview of this chapter. The persevering reader will of course continue reading.

Assume instance 1 has a value of 104/29/47 for the complexity predictor, and we wish to evaluate three other instances: 2, 3, and 4, to check whether they are harder than instance 1 or not. Such a hardness check requires solving the instance with the priority rules and composite branch-and-bound procedure to get a value for the complexity predictor for each instance, and this will be done in a stepwise approach as shown in the figure. First, each instance is solved using the quick and easy priority rules

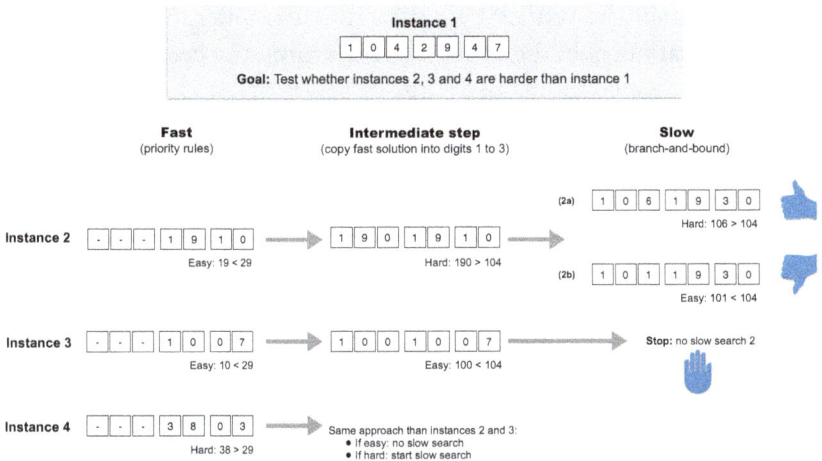

Fig. 5.3 Complexity predictor (example)

and lower bounds to find the values for digits 4 and 5 of the complexity predictor (Phase 1). Assume that this results in the following value - - - /19/10 for instance 2 indicating that no value is available yet for digits 1 to 3, and a 19% deviation is found between the lower bound and upper bound after 0.01 seconds. Based on this value, instance 2 looks easier than instance 1 since $19 < 29$, and so no slow search is required, saving computational time. However, the complexity predictor is not complete, and the first 3 digits should be filled out before a final evaluation can be made, but this typically requires more computational time. Therefore, in a second intermediate step, the 19 is copied into the first three digits of the complexity predictor (multiplied by 10), resulting in a value of 190/19/10. Such a copy of the percentage deviation found by the fast priority rule search to the three digits of the slow branch-and-bound search (intermediate step in the figure) can always be done since the real percentage deviation of the slow search will always be smaller than or equal to the percentage deviation found by the fast search. Consequently, the copy-paste shows a worst-case behavior as if the slow and intense search will not find any improvements. In the example, the intermediate value for instance 2 is assumed to be harder than instance 1 ($190 > 104$), and so this should be verified by starting the time-intensive search (slow) to find the real values of digits 1 to 3 of the complexity predictor. Suppose that this slow search results in a value of 106/19/30, and then instance 2 (2a) is indeed assumed to be harder than instance 1. Of course, if the slow search results in a lower value for the complexity predictor, then it confirms that instance 2 (2b) is indeed easier than instance 1, and the slow and time-consuming search was not necessary. This approach is used to save computational time for some instances, as illustrated by instance 3. After the quick and easy search of digits 4 and 5, instance 3 can be classified as easier than instance 1 (10 < 29), and there is no need to start a slow branch-and-bound search since the intermediate step confirms that the instance is indeed easier (100 < 104). Of course, when the fast search results in a harder instance, as is the case for instance 4 (38 > 29), this should be confirmed by following the same logic as instance 2.

Are you still following along? Maybe you should read the example again.

If not, it suffices to understand that José and I were thrilled with this complexity predictor because suddenly, the path was cleared to subject many projects to our numerous modifications, which was ultimately the goal of this study.

5.4 Small Projects Are Not Easy

In the end, after much testing, we achieved what we aimed for and obtained 623 small but difficult projects. This experiment was the most intensive one we had conducted up to that point. I previously mentioned that it took more than 40 years in total to execute the three phases of the project, but this is actually an understatement. The truth is that these 40 years represented the duration of our final experiment, but we had conducted dozens of other experiments before, often with equally long computational times, which led to nothing.

Perhaps that is precisely why both José and I love research so much. It is a quest for a clearly defined goal (finding small complex projects), but it is mainly a journey to find the path to reach this goal. Our three-phased procedure with the five modifications, including the use of the complexity

5 Why Is the RCPSP So Difficult? (Part 1)

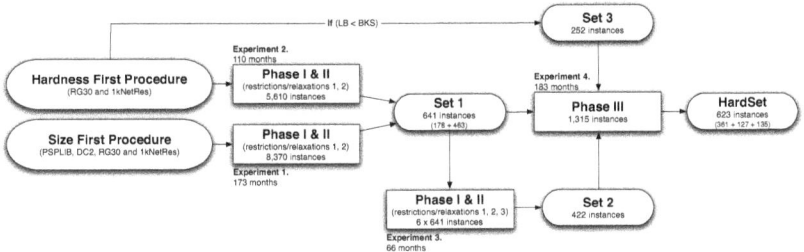

Fig. 5.4 Going to the core (Part 1)

predictor in Phase 1, eventually looked after many failed experiments like the flowchart in Fig. 5.4. Only after reaching this point, could we initiate our final experiments on the supercomputer. I will not go too much into detail of the various elements in the figure, but the attentive reader will notice the 110 + 173 + 66 + 183 = 532 months of computer time, which corresponds to 44 years.

We compiled the 623 new instances into a new dataset called the CV dataset. Apparently, it looks like we were as exhausted as the supercomputer and had lost our creativity, as we simply used the first letter of our last names.[3] We hoped that these project instances would provide motivation for researchers to try to solve them optimally. We could not solve them optimally ourselves, no matter how hard we tried with many algorithms from the literature. Perhaps researchers, being as creative as they are, would come up with innovative algorithms to demonstrate that they could search for solutions in a completely new way, thereby solving these projects. We hoped for innovation, for new insights, for additional algorithms, and for progress that would disrupt current standards. At the end of our published paper, which I mentioned at the beginning of this chapter, we wrote the following words:

> We believe that this research can stimulate future research in different ways, and we certainly hope that it will lead to radical breakthroughs rather than incremental improvements when developing new algorithms. In the past

---

[3] If you do not understand what I mean, it is C = Coelho and V = Vanhoucke.

decades, a lot of new algorithms have been proposed to solve the RCPSP, and while many of them certainly have inspired researchers to create new insights and understanding, some of them merely ruminated existing ideas to present only minor improvements. We believe that trying to solve the unsolvable instances will require a fundamentally different approach to find better solutions, rather than just small improvements.

And so, by the end of 2019, we had completed yet another study, which we once again submitted to the *Computers and Operations Research* journal, where it was published one year later after several revisions. Research is a time-consuming pursuit, but the satisfaction of seeing your article appear in a professional journal continues to make it worthwhile.

Only recently, a few months before I started writing this book, I asked José which study he was most proud of, and he said without hesitation that it was our *"going to the core"* study in this chapter. I could not agree more. Perhaps it was the intensity of the experiments or the endless discussions during the summers in Belgium that made it so rewarding. However, I believe that the idea of searching for small but difficult projects was both simple and original, containing a certain beauty that we cannot fully explain. Or perhaps it was simply the prospect of encouraging future researchers to develop better algorithms for our favorite project scheduling problem that gave us hope.

Not long after, we received messages with proven optimal solutions for some of the CV instances.

It is incredible how many hardworking, but especially creative, researchers there are in the project scheduling domain.

6

Can We Make the Projects Easier? (Part 1)

> **Note:** This chapter is based on the article "*An analysis of network and resource indicators for resource-constrained project scheduling problem instances*", published in *Computers and Operations Research*.

Immediately after the submission of the paper of the previous chapter, we already had new ideas for additional research. In 2019, our paper was still under submission, and José and I often consider the period between submission and reviewer feedback as a very pleasant period. It means that we can still fully enjoy our achieved results, which we are always very satisfied with, without yet worrying about the additional work the reviewers will give us. During this period without worries, we can do little to nothing on our existing research, and that is usually the perfect moment to think about new ideas.

And that is exactly what we did.

In our pursuit of complex projects from the previous chapter, we noticed something intriguing. Each time we made a modification to a project, the values of the network and resource indicators naturally changed. These changes in values were logically explicable, as we not only removed activities from the project networks but also altered the resource constraints. However, as we mentioned in the previous chapter, we frequently observed that the projects did not become more difficult at all after these modifications, but instead, they became most of the times much easier to solve. This occurred mostly when activities were removed from the project, making the new project smaller and thereby easier to solve. However, what we did not fully comprehend was that for the projects where only the resources were changed, and not their network, some of them became easier to solve, while others became much more complex. We realized that we did not fully understand the effect of the resource changes.

To compound our lack of understanding, upon closer examination of the projects with increased complexity, we occasionally discovered that both the unchanged and modified versions actually had the same optimal schedule. However, while this schedule was easily found for the unchanged project, it posed a considerably greater challenge for the modified one, exacerbating our perplexity. So we had two different projects, both identical in the project network but slightly different in the resource parameters, yet both with the same optimal schedule. Since we aimed to find optimal schedules for projects, it seemed better to us to use the easy version of the project in our scheduling algorithm, so that we could find its optimum much more quickly than when using the complex project. An interesting thought, but then we must, of course, know how to distinguish between an easy and a difficult version of the same project, and we must be certain that the schedules of both projects will be the same. And for the time being, we still did not know that for sure, but both José and I found that thought, at the very least, very interesting.

We began sharing some ideas via email, which is how every research study of ours starts, and gradually started to contemplate how we could modify projects anew, much like in our previous study but with a twist. While our previous study was a controlled search for more complex projects, we now planned to work in the opposite direction, making changes to projects to simplify them without altering their schedules. After several months of intense idea exchanges, we were ready in early 2020 to meet again in Lisbon to shape our new idea.

And then came Covid-19.

6.1 Solution Equivalency

Research is not a job you easily do online because it requires a kind of mutual understanding where collaborating partners consider each idea valuable until a better one emerges. Such a process of idea generation and refinement, which is then repeatedly assessed, adjusted, and improved until the best idea surfaces, demands ongoing collaboration, ideally done on paper and in close proximity. But then came the coronavirus, suddenly taking us by surprise, forcing us to try it differently. I had never really been fond of Skype, and rarely used it, but when Zoom emerged at that time, both José and I knew that we could explore further possibilities with it. That is how our very first remote collaboration began.

Our quest for modifications in projects to make them easier (as in this chapter) or even more challenging (as in the previous chapter) has always had the same goal in mind. Through these modified projects, we aimed to gain insights into the complexity of projects so that scheduling algorithms could be better tailored to certain project characteristics, making the search for an optimal schedule as fast as possible. The reason why the resource-constrained scheduling problem is so difficult for many projects is known and lies in the fact that there are too many feasible schedules (solutions), making the search space for an algorithm immensely vast. Thus, every algorithm's search must navigate cleverly through this search space without blindly enumerating all possible schedules in the hope of quickly finding the optimal solution. Since the search for each algorithm proceeds differently, it is entirely possible that one algorithm works better for one project, while another algorithm is better for other projects. Therefore, selecting the fastest algorithm for a particular project could improve this search, and this selection likely depends on several characteristics of a project measured by the known network and resource indicators.

In Chap. 1, we had already conducted a lot of research on the network indicator, but we had somewhat neglected the resource indicators. This time, we would put them in the spotlight.

The previously referenced research on phase transitions, which denotes a significant change in the behavior of a computational problem as certain parameters are varied, was highly relevant to our new study. Remember that such transitions indicate critical points where the scheduling problem becomes significantly harder or easier to solve, or where the properties of the solutions change abruptly. The importance of resource indicators had already been extensively studied in the literature, and despite the lack of consensus, it was known that for some values of these indicators, the RCPSP was more easily solvable than for others. For instance, a certain study showed that it could be much more difficult to find optimal schedules for projects with resource-constrained (RC) values around 0.4 than for projects with much lower or higher values. In other words, the RC indicator exhibits easy/hard/easy transitions, with complexity increasing significantly as projects approach the RC value of 0.4. Similar insights were also known for other resource indicators, such as resource strength, although there was still some debate about the correct values of the indicators for the observed transitions.

It is precisely these debates that formed the basis of our new study.

The existence of phase transitions means that there are zones of "*easy*" and "*difficult*" projects defined by the values of these resource indicators. However, if we would be able to modify projects from the difficult zone (e.g., around 0.4 for RC) in such a way that they move into the easy zone (with RC values closer to 0 or closer to 1) without changing the set of possible schedules, then perhaps the discussion about the correct values of the phase transitions would be entirely justified. I want to clarify immediately that we had no intention of questioning the entire concept of phase transitions, but rather wanted to demonstrate that caution is necessary when using certain resource indicator values to predict the complexity of a project. If these values can change without altering the project and its underlying number of schedules, then they are not very reliable and should be questioned.

After many online meetings working on this idea, it ultimately led us to a new concept that we called "*solution equivalency.*" Two projects with the same network structure but different values for the resource parameters are solution equivalent if it can be demonstrated that their entire set of all possible feasible schedules are identical. This definition implies that both projects must consist of the same set of activities with

6 Can We Make the Projects Easier? (Part 1)

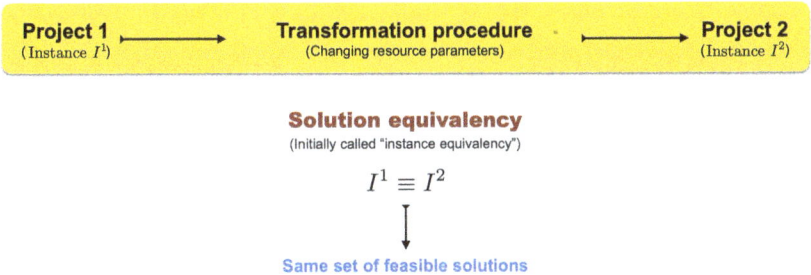

Fig. 6.1 Equivalent projects (changing resources)

the same precedence relationships between them, but their resource usage and availability can be completely different. Usually, the possible project schedules will look very different due to these differences, but if in a rare case they do not, then the two projects are called solution equivalent. Because of these identical schedules, the total search space of all possible schedules is also equal, and thus, the projects have equal complexity, even though the resource indicator values are completely different due to their different resource data.

Because stumbling upon solution equivalent projects by chance is not very likely, we decided to generate them ourselves by modifying existing projects as visualized in Fig. 6.1.[1] More specifically, we chose to alter the resource parameters of an existing project in such a way that not a single schedule was eliminated from its solution space, thus ensuring solution equivalence.

The new concept of solution equivalence thus consists of changing a project instance I^1 to a new instance I^2 without altering the search space. For this, instances I^1 and I^2 must surely have the same network (i.e., the same number of activities and precedence relations) because changes in these parameters make the projects entirely different (and thus the search space as well). But it is possible to change the values of the resource data by assigning more or less resources to the activities without seeing much

---

[1] In our original study referenced at the beginning of this chapter, this concept was called "*instance equivalency*," but we later changed it to "*solution equivalency*" after discovering new forms of equivalence presented in Chap. 8.

difference between them. If both projects still have the same possible set of solutions (schedules), and thus the same optimal solution(s), it could very well be that one project lies in the hard zone, while the other lies in the easy zone, because they have different values for the resource indicators, even though both projects are the same. This makes the resource indicators not very reliable, and choosing the right project (with values for the resource indicators) is crucial to fully exploit the potential of possible phase transitions.

With each modification, it was forbidden to adjust the number of activities and their durations (which was possible in the previous study in Steps 1 and 4), and the precedence relations between them had to remain unchanged too. The resource parameters, however, could be modified. But unlike such similar changes in our previous study, they had to be approached very carefully this time because we always had to ensure that the modified project remained solution equivalent to the version before the modifications.

Fortunately, we had enough projects available from our previous studies, so we only needed to search for rules to implement these modifications in the way that we wanted. We did this by using four new theorems that we incorporated into a completely revamped transformation procedure.

6.2 Changing Resources

Despite our experience in modifying projects in the previous studies, this time, we had to proceed more cautiously. Previously, we made our changes more or less thoughtless using the five discussed steps, and with each change, we experimentally assessed whether the project became easier or harder. Each modification could lead to a new and entirely different project, and the concept of solution equivalence was not born yet.

This time, the changes had to be executed very precisely as we had to ensure that no possible solution (schedule) was eliminated from the solution space with each modification. This means that no schedules could be added or removed when the modifications are made to instance 1 that lead to instance 2. Every possible feasible schedule from instance 1 had to also be feasible for instance 2 and vice versa. To guarantee this, we defined four theorems that we incorporated into a new transformation procedure to change projects into new solution equivalent projects. These theorems were described mathematically in the study (including a proof for their solution equivalence) and are described below in a general manner without mathematical details:

Theorem 1 (Irrelevant Resource Demand) When the demand for a particular resource for an activity is so small that it can never lead to a resource conflict, that resource demand can be set to zero, as if that activity uses nothing of this resource at all. This happens when the combined resource consumption of the activity in question and all other activities that can be scheduled concurrently (due to the absence of precedence relations) is lower than the resource availability. By setting this resource usage to zero, the resource indicator values of the project will naturally be adjusted, yet the set of possible solutions will remain unchanged.

Theorem 2 (Dominated Resource) A project can obviously have multiple types of resources, each with its own availability and activity requirements, each potentially causing resource conflicts. Under certain conditions, though, one resource type may be entirely dominated by another, meaning that every conflict involving one resource also involves a conflict in the other. In such cases, the dominated resource can be completely removed while preserving the potential conflicts of the retained resource and maintaining the same set of schedules in the project.

Theorem 3 (Join Resources) Alternatively, when two resource types may not dominate each other (and thus cannot be removed), they can sometimes be merged into a single new resource. This happens when all activities that can be scheduled concurrently use only one of the two resources, which are then consolidated into a single combined resource. This results in the new project having one less resource, leading to new values for the resource indicators without changing the set of feasible schedules.

The omission of mathematical details for these theorems entails a risk that the conditions to apply these theorems may not be entirely clear.

Therefore, without delving into specifics, I would like to reiterate that each of these theorems imposes conditions on a set of activities that can be concurrently scheduled according to the network. Activities that meet this criterion are referred to as "*precedence-compatible*" activities, and determining compatibility between two activities is straightforward. If one activity has no predecessor or successor to the other in the project network, then they are considered precedence compatible. Such a criterion can be efficiently checked for all activities in a project network, requiring no significant time or effort for a computer to perform rapidly, even if this process needs to be iterated thousands of times consecutively.

But for Theorem 4, this was by no means the case.

The reason is that just because activities are precedence compatible does not guarantee they can be scheduled together. If their total resource consumption is lower than or equal to the resource availability, then there is no significant issue as Theorem 1 could be applied. However, it is also possible that their total resource consumption exceeds the resource availability, leading to a resource conflict. In such cases, these precedence-feasible activities are not "*resource compatible.*" It is known for the RCPSP that resource conflicts, resulting from such resource incompatibility, can be resolved in multiple ways by scheduling one or more of these resource-incompatible activities earlier or later in the schedule. This complexity is precisely why the RCPSP is classified as NP-hard.

The conditions for Theorem 4 are based on enumerating all possible ways to resolve these resource conflicts for resource-incompatible activities, and this number can quickly escalate to a very large figure. Specifically, the theorem will enumerate all (precedence and resource) compatible sets for each activity of the project, and each of these sets, being resource compatible, will never exceed the resource availability. However, when the maximum total resource consumption of all sets is lower than the resource availability, it implies that there is actually a resource surplus that will never be utilized. We called this surplus the resource waste, leading to our fourth theorem:

Theorem 4 (Resource Waste Procedures) If the maximum resource consumption of all sets of compatible activities is lower than the resource availability (resource waste), then the availability of this resource can

6 Can We Make the Projects Easier? (Part 1)

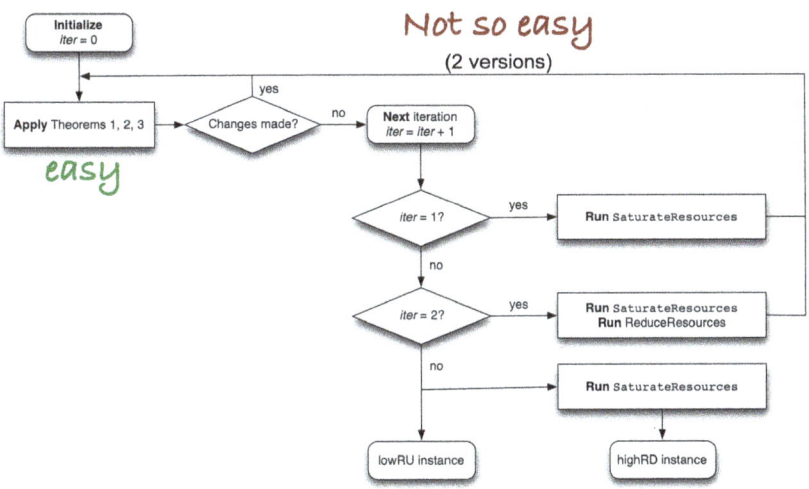

Fig. 6.2 Transforming projects (Part 1)

be reduced without changing the set of compatible activities, and thus, without eliminating a possible schedule from the project.

Since the number of compatible sets of activities of the project can often be substantial, Theorem 4 is considerably more challenging than the previous three theorems and can lead to significant computation times. Therefore, this fourth theorem was incorporated into a transformation procedure visually depicted in Fig. 6.2, where the easy first three theorems are applied first before delving into the challenging quest for the fourth theorem. Moreover, Theorem 4 could be applied in two ways. When a positive resource waste value was identified, the resource requirement of some activities could be increased (since there is a surplus) to saturate the total resource consumption. Such adjustments led to modified projects where the resource demand is maximized, which we ultimately termed as the highRD instances. Additionally, this resource waste could also be used, with some additional conditions, to set the resource usage for certain activities to zero. This modification ultimately resulted in modified projects that we referred to as the lowRU instances. Thus, after a series of modifications based on the four theorems, each project could lead to two new modified projects (lowRU and highRD) with significantly different values for the resource indicators, but both having the same number of feasible schedules as the original project.

Each project could thus lead to three solution equivalent projects.

We used this procedure for 10,793 of the 15,373 projects we had up to that point, resulting in three times as many projects. From now on, each project had two identical twins. Researchers could now choose from three versions of the same project before starting the search for an optimal solution. And since we know that each of the three projects has the same set of solutions, the choice can be made completely arbitrary.

Or maybe not.

6.3 Reliable Resource Indicator

Since the three twins had different values for the resource indicators but all shared the same set of feasible schedules, it was futile to investigate whether one twin would be easier to solve than another. When the search space of possible schedules is equal, the complexity is actually the same, and there can be no notion of difficult or easy versions of the same project. This seemed logical, but nonetheless, we conducted the test.

You never know in academic research, and secretly, we hoped that we would be proven wrong.

To ensure the validity of our hope, we used both the 48 branch-and-bound procedures and the 3 meta-heuristics that we also used in the previous chapter to solve the three twins out of the 10,793 projects. Thus, we had a total of $3 \times 10,793 = 32,379$ projects to examine whether a particular twin—the original, lowRU, or highRD project—would be more difficult to solve than the others. However, we were unable to observe any significant differences, completely dashing our faint hopes of discerning variations in complexity.

It was, of course, somewhat naive to hope that an algorithm would make a difference in exploring the search spaces of the three twins because they were simply not different at all. However, we soon realized that we should not be searching for the complexity of the twin projects, but

6 Can We Make the Projects Easier? (Part 1)

rather for the predictability of this complexity. The definition of phase transitions taught us that the resource indicators must have sufficient discriminative power to indicate when a project would be easy or difficult to solve. Instead of looking for easy or complex projects, we needed to search for the ability of the resource indicators to predict this difference in complexity. It may seem like a small difference, but it turned out to be crucial in validating our new transformation procedure.

The fact that the three twins of each project share equal complexity now became the starting point to test the predictive power of four different resource indicators. It mattered little whether a project fell into the easy or difficult zone of a resource indicator; rather, it was crucial whether a particular resource indicator could recognize this. If the three twins have equal complexity, whether easy or difficult, then the values for the resource indicators should also be approximately equal, placing the three twins in the same easy or difficult zone. However, when the values of the resource indicators for the twins vary greatly, one project may fall into the easy zone, while another falls into the difficult zone (or vice versa), despite having equal complexity, significantly reducing the relevance and reliability of the resource indicator as a complexity indicator.

Stable values for the resource indicators indicate better predictability of project complexity.

Therefore, we first and foremost examined the values of the four resource indicators RF, RU, RS, and RC that we introduced earlier.[2] Their average values were indeed significantly different between the three versions of the 10,793 projects, as shown in Fig. 6.3. It initially seemed tempting to conclude from this that none of the resource indicators is reliable. However, such a conclusion would be too simplistic, and thus entirely incorrect, as these values are only averages for the 32,379 projects. Nonetheless, the difference in average values indicated that the existing knowledge of the phase transitions must be interpreted with caution. This gave us hope and motivation to explore our inquiry more deeply.

To assess the stability of the resource indicators, we compared the three versions of each project with each other instead of simply looking at

---

[2] These four resource indicators were presented in Fig. 2.1 of Chap. 2.

	RF	RU	RS	RC
Original	0.70	3.01	0.29	0.43
lowRU	0.55	1.82	0.25	0.55
highRD	0.77	2.52	0.17	0.57

Fig. 6.3 Resource indicator values

average values across all projects. Additionally, since we were particularly interested in the predictive power of complex projects, we decided to determine whether each project was easy or difficult to solve. We did this in two ways. Firstly, if a project could be optimally solved using our branch-and-bound procedure of Chap. 4, even if it took a very long time, we classified the project as easy. If not, we classified the project as difficult. A similar approach was taken using the three meta-heuristics, but since this method cannot guarantee optimality, a different criterion was used to classify a project as easy or difficult. Specifically, we looked at the number of schedules that needed to be generated to find a good, but not necessarily optimal solution. If this number was low, we considered the project easy; otherwise, we considered it difficult.

With this approach, a project could be considered easy for a meta-heuristic but difficult for the branch-and-bound procedure, and so it is important to understand that this classification does not represent a theoretically correct categorization of project complexity. Nevertheless, our experimental approach was sufficient to begin our study on the stability of the resource indicators to measure their power to predict project complexity.

We divided the values for each resource indicator into different equally sized intervals between their minimum and maximum values and examined how many projects fell into each interval. Then, we looked at the percentage of difficult projects in each interval (as defined earlier) to identify the interval with the most complex projects. It was, for example, expected that there would be many more complex projects in the interval 0.35–0.45 for the RC indicator than in the interval from, say 0.75–0.85.

Meta-heuristics					Branch-and-bound				
	RC	RS	RF	RU		RC	RS	RF	RU
Original	2.6%	3.1%	1.5%	1.8%	Original	6.8%	9.3%	6.7%	5.9%
lowRU	3.4%	3.0%	4.5%	5.4%	lowRU	11.2%	8.5%	12.9%	14.7%
highRD	3.9%	3.0%	2.3%	3.2%	highRD	11.3%	4.3%	4.4%	9.6%

Fig. 6.4 Clusters for resource indicators

Finally, we examined the variability in the percentage of complex projects in each interval. We concluded that resource indicators with almost equal percentages of complex projects in all intervals are not very discriminative, as all of their possible values have an equal chance of containing complex projects. Such indicator would have a very low variability value. However, when the variability for another resource indicator varies greatly, it means that certain intervals contain a lot of complex projects, while others have hardly any, indicating a good potential to discriminate between easy and hard projects and thereby possessing a high predictive power for project complexity.

The variability is shown in Fig. 6.4 for the four resource indicators, each split for the three twins of the projects (original, lowRU, and highRD) and for the two procedures (meta-heuristics and branch-and-bound). First and foremost, we noticed that the variability values were higher for the exact methods than for the meta-heuristic methods, indicating that the resource indicators could be better used to predict the project complexity for exact methods than for meta-heuristic methods. This is logical, of course, since meta-heuristic methods explore the search space in a more or less controlled random manner to find a good schedule, introducing an element of luck, while the exact methods have to explore the search space systematically to guarantee optimality.

But what particularly excited us was that the predictive power of the resource indicators was clearly higher for the modified instances (lowRU and highRD) than for the original projects, except for the resource strength. This indicates that the complexity of the modified projects might be more predictable compared to the original projects, at least for three of the four resource indicators. Moreover, since the modified projects were obtained by removing all unnecessary resource data using the four theorems, this probably also means that these projects are "purer" and have a resource structure that lies closer to the "true" values for the resource indicators.

In our enthusiasm, we recommended in the conclusion section of our study to researchers to always test their new scheduling algorithms on both the original and the modified instances. We were convinced that our results had shown that the modified instances could better reveal phase transitions, which could lead to better insights and a better tuning of the algorithms to the characteristics of the project.

We had created three times as many projects. Our new projects were not easier to solve, but at least they were more predictable.

6.4 Resource Strength

It would not be entirely fair to mention the disappointing results of the resource strength without discussing it. Disappointing might be the wrong word, as the results just showed that the predictive power decreased with modified projects, indicating that it was most reliable for the original projects. That observation was not very clear for the meta-heuristics, but it was noticeable for the branch-and-bound procedure, and I prefer to call it a pleasant observation rather than a disappointment, even though our modifications could not increase the predictive power of that resource indicator.

I call it pleasant because this resource indicator had received quite a bit of criticism in the literature for not being a pure resource indicator. That is completely justified because the RS formula not only measures the resource scarceness but also includes the network structure. This means that this indicator's predictive power is not only determined by the resource values of the projects but also by their type of network. What I find particularly surprising, pleasantly surprising even, is that the authors who developed the PSPLIB focused the project's resource data primarily on that RS parameter (they also used the RF parameter, which is less reliable). I find it incredible that they made the right choice in building the resource data around the RS indicator. Perhaps they had already realized in 1995 what we were only discovering in our study more than 20 years later, or perhaps that choice was purely coincidental.

I think it was a kind of gut feeling, and I know that it is wise not to underestimate researchers' intuition too quickly.

After a year and a half of online work, we submitted our paper to the *Computers and Operations Research* journal during the summer of 2021. It was accepted less than one year later after undergoing some minor revisions. Throughout our research period, we had not met in person even once. The entirety of our work was completed only through digital means. We had become masters of online meetings, able to take notes and brainstorm as if we were sitting next to each other using an iPad and shared screen. However, it was not the same feeling as our real brainstorming sessions, and the chemistry between us was missing.

But there was hope. The coronavirus showed signs of weakness and began to gradually diminish in strength after a year and a half of tyranny.

But not us. Both José and I still felt as powerful as if we were in our early thirties when we started our collaboration.[3] With a liberating feeling after the long imprisonment, we enthusiastically embraced our renewed freedom and promised each other to meet again as soon as possible to start our new research idea. The feeling of excitement about returning to new research reminded us of the words of Al Pacino in the film "Sea of Love."

We felt like f*** teenagers.

---

[3] As a matter of fact, when we first met in April 2002, José had just turned 30 the month before, and I, also in March, had just started my last year in my third decade.

7

Why Is the RCPSP So Difficult? (Part 2)

Note: This chapter is based on the article "*New resource-constrained project scheduling instances for testing (meta-)heuristic scheduling algorithms*", published in *Computers and Operations Research*.

In the fall of 2021, just after the submission of our previous study, we received the relieving news that travel was permitted again. It was a very busy period at our universities as we transitioned to a new post-Covid-19 work mode, leaving little time for our research. Fortunately, José made use of every free day or extended weekend to travel to Ghent to discuss ideas in person. With renewed enthusiasm and joy, we gathered many ideas to start new research projects.

It seemed like Covid-19 was finally over.

During José's short visits to Ghent, we started to pick up the two previous studies again, to view them from a renewed perspective, and to further develop them into new research ideas. In these previous studies, we mainly focused on examining the complexity of projects to find optimal schedules for the RCPSP. We primarily relied on our 48 different branch-and-bound methods from Chap. 4 because they guarantee optimality, even though we were aware that such a quest for optimality quickly reaches its limits. Beyond a certain project size, the pursuit of optimality simply becomes impractical, even when having access to a very powerful supercomputer that José has become so enthusiastic about over the years. It is precisely for this reason that in Chap. 5, we deliberately kept the new projects from the CV set very small. Our goal was to generate small yet challenging projects in the hope that they might eventually be optimally solved. Not by us, but perhaps by others. We had been able to demonstrate that for these projects, optimal solutions could not be found with our exact branch-and-bound method, even after long computation times, labeling these projects as highly complex. But the underlying idea was that other researchers should be able to find optimal solutions to those projects with enough creativity.

In the academic world, projects are mostly very small, and sometimes very challenging.

Despite our hope that for these small projects optimality could eventually be found with sufficient computer power and a clever new algorithm, for some large projects it remains simply impossible to find such an optimal schedule, and we must settle for a schedule that approaches optimality. Meta-heuristic algorithms, which also have received a lot of attention in academic literature in recent decades, are highly suitable for solving such large projects to near optimality. That is why these methods are much more popular in business than our branch-and-bound procedures. They are fast and work pretty well, even though they no longer guarantee optimality.

Projects in the real world are often very large.

So, we began to worry that we might have been too one-sided in our search for small challenging projects to feed our quest for optimal schedules. Maybe we had not paid enough attention to the larger projects that can also be solved quite well with these meta-heuristics. We began to wonder if our definitions of complexity of projects would still hold true if we no longer wanted to guarantee optimality but were satisfied with a reasonably good solution.

And immediately came the question of what "reasonably good" actually means.

Asking the question is trying to answer it. That is the mission of academic research. Luckily, the winter break of 2021 was approaching, providing some free time. Since neither of us is a big fan of the Christmas atmosphere, I booked a ticket to Lisbon to use these holiday weeks to organize our loose ideas of the past few months. Without face masks or other restrictions, we had real-life discussions and brainstorming sessions. It felt wonderful, and we were convinced that our new study, after the dark period of Covid-19, would become one to remember.

We were back in research business as usual.

7.1 It Is No Luck

The reason why our branch-and-bound procedures could only be used to optimally solve relatively small projects is obvious. This method systematically explores the search space of all possible schedules by constructing a search tree.[1] This search tree is only gradually constructed and explored going deeper and then turning back, and this process continues until an optimal schedule is obtained. Even for medium-sized projects, such tree can become immeasurably large, taking up huge amounts of computer time. There is simply no other choice but to keep the projects sufficiently small for these exact methods.

But for meta-heuristics, this necessity for small projects is much less relevant. Meta-heuristics do not systematically search for the best schedule like exact methods do. They do not construct a search tree as exact methods do, but instead roam the search space in a random yet guided manner. For small projects, it is therefore quite possible for meta-heuristics to find the optimal schedule, but that is more of a coincidence because of the relatively small search space. However, for large projects with an immensely vast search space, it remains a challenge to find a very good schedule, and thus, the exploration in the solution space must be well-guided. It is especially for such projects that meta-heuristics become

---

[1] Remember that this search tree can be built in various ways using the five components briefly mentioned in Chap. 4.

interesting. When the search space becomes immeasurably large and exact algorithms will never be able to guarantee optimality, meta-heuristics can make a significant contribution by arriving at a very good, possibly the best schedule based on guided luck.

You need a portion of it, but it is better not to rely too much on luck. A well-crafted meta-heuristic should find a good solution in a structured way and should perform much better than a random walk in the solution space to find a schedule by coincidence. Some say you need a little bit of luck in love to find a partner for life. Others believe it is not luck at all, and if someone is destined for you, they will find you. But once you have found the right partner, how do you know whether it was pure luck or predestined destiny?

That question can also be asked for meta-heuristics.

How can you be certain that the success of a meta-heuristic search is not simply a matter of luck?

The answer lies in the creativity of the designer of the meta-heuristic, who cleverly combines various components to replace randomness with a highly targeted search for good solutions. If the meta-heuristic searches randomly, it is akin to trying to shoot a bird with your eyes closed. You might be lucky to hit one when enough birds are flying over your house, but when most of them are elsewhere, it is very likely you will miss them.

That is why no effective meta-heuristic performs its scheduling quest solely through random searches. A good meta-heuristic is a procedure where coincidence plays as small a role as possible, and which directs the search as well and quickly as possible toward the right direction (of the search space) using various components. Techniques such as crossover operators are used to combine the best elements from two schedules into a (hopefully) better schedule. A local search method is also occasionally applied, making small changes to obtain an even better schedule during the search, and mutation operators are often needed to jump to a completely different part of the search space if the algorithm thinks the current part is not promising enough. Developing a good meta-heuristic is an art in itself and has many more degrees of freedom than exact methods, with the main goal of transforming the luck factor into a fast and intelligent search for the correct region of the search space of all possible schedules.

It is evident that more than mere luck is necessary, and that became the major insight for our new study.

If you are wondering why I am writing about love and birds here, it is just to illustrate that this is how José and I communicate when generating our new ideas for project scheduling research. What holds true for birds also applies to projects. When projects are easy to solve, it might mean that there are enough relatively good schedules in the search space, and you do not need a well-crafted meta-heuristic to find one. Often, a simple random search might be sufficient, and you can count on a little bit of luck. But for complex projects, these reasonably good schedules are rare, or at least well hidden in corners of the search space (figuratively, of course), and the chance that you find one just by a random search decreases to such a low level that you need much more than just a little bit of luck.

I know that this insight does not sound very innovative, as it is well-known that randomness should be minimized as much as possible in a meta-heuristic. However, we nevertheless used that obvious idea of shooting birds to redefine the complexity of projects for meta-heuristics.

We called it sigma distance.

7.2 In Distance Lies Complexity

Because we were searching for a new definition to express the complexity of a project when meta-heuristics are used, we could no longer rely on the definition from Chap. 5, as it depended on *optimal* schedules and *exact* algorithms. However, with a slight modification, this definition could be restated as follows:

> **Definition**: A project instance is assumed to be hard if a reasonably good solution can not be found by a simple random search but instead requires a well-crafted state-of-the-art heuristic algorithm.

The definition clearly expresses that there must be a significant difference between randomly generated solutions and the solutions from a

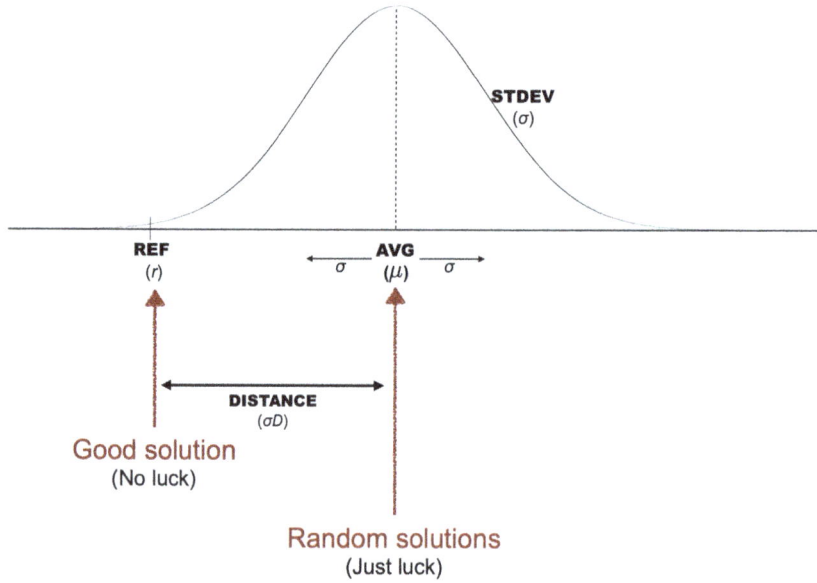

Fig. 7.1 Sigma distance concept

cleverly designed meta-heuristic, and that is exactly what the new sigma distance concept measures.

To calculate this sigma distance metric for a project, a large number of schedules must be randomly generated. Each of these schedules will have a total project duration (makespan) that can vary quite significantly due to the randomness. After a sufficient number of schedules have been generated, they can be plotted on a graph as depicted in Fig. 7.1, where the average value for the project makespan typically represents the value around which the generated schedules tend to cluster. Because many schedules will have smaller or larger values obtained purely by chance, the standard deviation of the makespans must also be calculated to measure the spread of the generated schedules around the average makespan value.

To assess whether these randomly generated schedules represent good schedules for the project, they need to be compared with a very good known schedule for the project. Therefore, we need another reference point, which represents the best possible schedule for the project that can be found by an exact algorithm such as the branch-and-bound procedures from our previous studies, or if available, the best known solution from the literature.

The distance between the average value of the makespan from the randomly generated schedules and the best known solution for the project then becomes the measure to express how difficult it is to obtain this reference solution, thus becoming our sigma distance metric. To express this metric in a relative unit rather than in a time unit (makespan), the distance is measured as the number of standard deviations between the two values. The difficult projects are therefore the ones with a large value for this σD metric, as the chance of finding the reference solution through random search is then small.

This was our simple idea that stemmed from our analogy of shooting birds.[2]

It is a fairly straightforward idea, but the accuracy of this new definition depends on whether or not the randomly generated schedules follow a normal distribution, as the figure shows. Indeed, to accurately measure the probability of finding a very good schedule (equal to or even better than the reference schedule), we rely on the basic rules of statistics that assume normality, simply because the σD metric measures the distance as the number of standard deviations. It sounds straightforward to assume that the randomly generated schedules for a project follow such a distribution (why should not they), but after many generations, we observed that for many projects, that assumption was simply not satisfied.

The problem was that even randomly generated schedules constituted only a fraction of the total search space, posing the risk of bias. For small projects, it was fairly easy to generate all possible schedules to find out whether the distribution had a normal shape or not. However, for larger projects, only a fraction of all possible schedules could be generated, which disrupted the entire process and gave us multiple headaches.

We decided to calculate the σD metric for 10,793 projects from our datasets, doing so by randomly generating 100,000 schedules for

---

[2] I do want to note here that I would never shoot a bird myself. I enjoy seeing those creatures roam freely in their natural habitat.

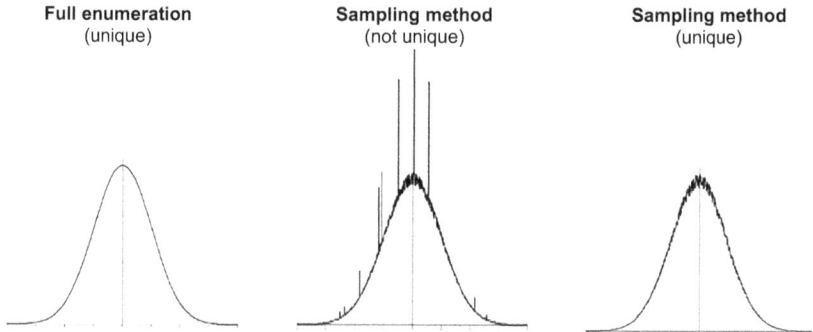

Fig. 7.2 Random sampling

each project in four different ways. Each schedule was represented by a randomly generated activity list that ranked the activities in a certain order, and we utilized both serial and parallel generation schemes to transform this list into a schedule. To maximize diversity in the schedules, we constructed them in both a forward (starting from the beginning of the list) and a backward (starting at the end) way, resulting in four schedules for each activity list.

Unfortunately, many of the generated schedules were not unique, and counting the duplicate schedules disrupted the shape of our graph, as shown in the middle graph of Fig. 7.2. To address this issue, we opted to remove identical schedules and only keep the unique ones, hoping that the graph would better resemble a normal distribution. While this adjustment improved the distribution somewhat, we still had to use a statistical normality test to verify whether the normal distribution assumptions applied. However, this was often not the case for many of our projects, leading us to remove numerous projects from our analysis. Fortunately, these removed projects were primarily the easy projects, while the challenging projects remained in our further analysis. After applying this generation and statistical testing process, we were left with only 2879 projects that could be used for further analysis of our new and promising concept. All others were simply discarded.

Not bad. But not perfect either.

7.3 Going to the Core (*Again*)

Even though we strongly believed in the σD metric as a measure of project complexity, we were somewhat disappointed that we could only calculate it for 26% (2879/10,793) of our projects. Furthermore, the values for this metric were not evenly distributed, so we could not draw many conclusions from our analysis. To truly test the validity of our concept, we knew that we would need projects with low, moderate, and very high values for the σD metric. Therefore, we concluded that we would need to search for new projects.

Fortunately, we had done it before.

In the "*Going to the core*" procedure of Chap. 5, we had already generated many projects that were difficult to solve for branch-and-bound procedures, resulting in the new CV set. We intended to use that experience and convert it into a new search for complex projects, this time for meta-heuristics. The flow chart of this generation process is shown in Fig. 7.3, and without going too deeply into every detail, its general operation is briefly described below.

We began with a large number of new projects with 50 activities generated by our RanGen network generator.[3] We performed various transformations on these projects to modify them, and with each small change, we checked if the σD value increased or not. It should not be surprising that this way of working was very computationally intensive, as the σD metric can only be measured after generating 100,000 random schedules, which had to be done after each change in the project. To make the process even more complex, we needed a reference schedule for each modified project. There was of course not a good solution available in the literature for these changed projects, since each change created a new one that never existed before. So we had to find a good schedule ourselves after every change, and we did that by running three existing and well-performing meta-heuristics from the literature until the best possible schedule was found, and called this the reference schedule.

---

[3] The attentive reader may recall that this project generator first appeared in our study in Chap. 1.

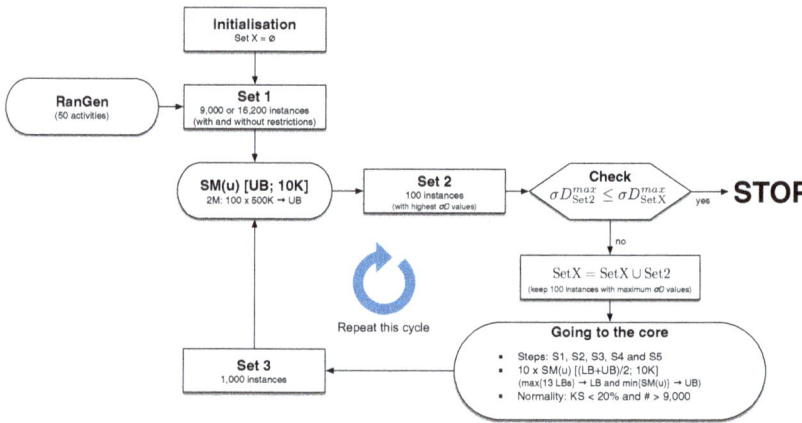

Fig. 7.3 Going to the core (part 2)

Despite the similarities of the new transformation procedure (Part 2) with the previous procedure of Chap. 5 (Part 1), it is the differences that deserve attention, as shown in Fig. 7.4.

Both procedures start with available projects, although the CV search of Chap. 5 was initiated with existing 30-activity projects from our database, while the new search was conducted on newly generated projects with 50 activities. The procedures for testing the project complexity were, of course, totally different, as we applied other definitions of project complexity. While our old transformation procedure mainly relied on our branch-and-bound procedures to determine whether an optimal schedule could be found or not, we now used a random sampling method and three meta-heuristics that we had developed in older studies to calculate the σD distance. Each of these three meta-heuristics used a completely different approach to arrive at good solutions and was published in renowned journals after an intensive review process, which led to the conclusion that they performed quite well. The first procedure was the genetic algorithm that we had already used in previous studies, but we also relied on a so-called scatter search method and an electromagnetic search method to find the best possible reference schedule. Lastly, the five different steps to make gradual changes to the projects were also somewhat different, although the difference was not very significant. Only two of the five

	Going to the core (CV set)	**Updated going to the core** (sD set)
Start	Existing instances 10,793	**New instances** 16,200 unrestricted and 9,000 restricted
Size	≤ 30	50
Complexity	Optimal schedules (Unsolvable projects)	Random schedules (High σD values)
Algorithm	Branch-and-bound (48 configurations)	GA, SS and EM (100 x 500k schedules)
Steps	Step 1. Remove activities Step 2. Remove resources Step 3. Change resource availability Step 4. Change activity duration Step 5. Change resource demand	Step 1. Remove precedence relations Step 2. Add precedence relations Step 3. id. Step 4. id. Step 5. Id.
Set	CV set (623 instances)	sD set (390 instances)

Fig. 7.4 Going to the core (comparison)

steps to modify the projects were adjusted to guide the search in the right direction. All in all, the differences between the two procedures are details, important to know if you were to implement them yourself, but not very relevant to understanding the rest of this chapter. From the original study, we were ultimately able to retain 623 projects, which we included in the CV set. In the new study, we obtained 390 new projects, which we introduced into the literature as the sD set.

This entire process took a lot of time, even with the supercomputer, and had to be restarted multiple times. The winter break was over, and I had to return to Belgium for teaching. However, we used the first two months of 2022 to run all our experiments and discuss the results over a video meeting. Since we wanted to obtain as many new diverse projects as possible with different σD values, we ultimately used 17,100 new projects under all restrictions, for which thousands of modifications were made. After long computation times, equivalent to years' worth on a standalone computer, we obtained many projects with σD values ranging between 3 and 18, resulting in 3100 instances in total. That was already very promising, considering that the existing projects from our database had much smaller values. For the existing PSPLIB projects, for example, the maximum σD value was slightly smaller than 8. However, despite these large values for our new projects, it did not necessarily mean that they were also more difficult to solve for meta-heuristics. And that was ultimately the main goal of our research project.

We, therefore, decided to analyze each project in detail to determine if there was indeed a relationship between a project's σD value and the quality of the schedule found with our three meta-heuristics. The search of these three meta-heuristics was terminated after finding 5000 schedules, and the best-found schedule was retained. To measure the quality of these schedules, we needed to know how far they were from the best possible solution for each project, and the smaller that difference, the easier the project was considered. Of course, this best possible schedule was already available in the literature for the existing PSPLIB projects, but for our new projects, it was not known at all. To obtain it, we once again used our three meta-heuristics, but this time we let them compute for a long time, generating up to 500,000 obtained schedules repeated 100 times, ensuring that it would be very difficult to find a better schedule.

The results are shown for both the existing PSPLIB projects (left) and the newly generated projects (right) in Fig. 7.5. The x-axis shows the value of the σD metric, while the y-axis represents the quality of the meta-

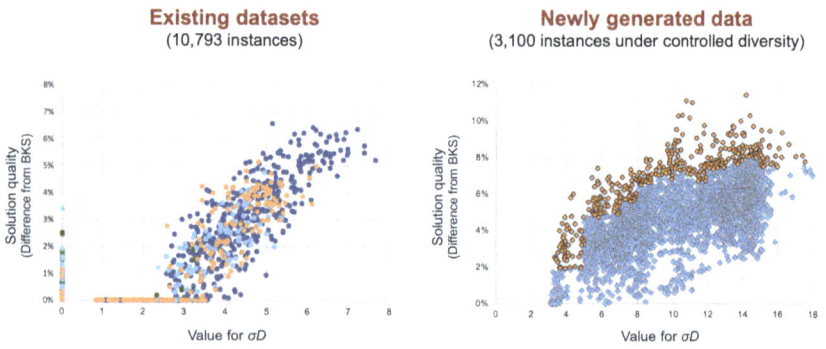

Fig. 7.5 σD values for two sets

heuristic solution. Low values on the y-axis indicate easy projects, while high values indicate greater project complexity.

Eureka!

It does not require a magnifying glass to see that the graph clearly shows a positive relationship between the σD value and the difficulty of a project. This relationship is very clear for the PSPLIB instances, but also for the new projects, which have much higher σD values, there is a clear positive trend. Our σD metric is thus clearly capable of representing the complexity of a project when the goal is to find a reasonably good, but not necessarily optimal schedule.

Another breakthrough in our eternal quest for insights into the scheduling problem that has been occupying us for 20 years now.

Perhaps there is such a thing as luck after all.

7.4 New Dataset (*Again*)

With these promising results, it was time to kick off the final sprint, which means writing and discussing to compile everything into a beautiful paper. Both of us had experienced during Covid-19 that this process does not go very smoothly from a distance, so I decided to return to Lisbon again. It was April 2022, and I quickly booked a plane ticket to leave the next morning. Apparently, I was not the only one eager to travel after two years of corona. Everything was fully booked, and I was forced to choose an apartment across the Tagus River, in Cacilhas in the municipality of Almada. I took the boat to Cais do Sodré every day and then walked to Jardim do Príncipe Real[4] with one goal in mind: We had a total of three weeks to analyze all our results, structure them, and decide how and what to write down in our new paper.

While writing, we came up with the idea of selecting a set of our new projects that we could make available to researchers. We chose 30 projects for each value of σD from our generated projects, ranging from σD of

---

[4] Jardim do Príncipe Real is the place where we had worked several times a week during my sabbatical on the study from Chap. 4 a few years earlier.

3 (easy projects) to σD of 15+ (very difficult projects). The selection of these 390 projects is shown in the top colored projects of Fig. 7.5, and we included them in our new set, which we called the sD set (with a small *s* and a capital *D*. I do not know why).

And so we had again a set of new projects that ranged from easy to difficult, but not for the exact methods this time, like the CV set, but now for meta-heuristic methods. We hoped, just like with the CV set, that researchers would be encouraged to develop new procedures, meta-heuristics this time, and advised them to conduct their tests especially on the complex projects with high σD values. We presented this set in our new paper, which we ultimately submitted to the *Computers and Operations Research* journal (yes, again) in April 2022, where it was accepted after two revisions in 2023.

Our first research project after the dark Covid-19 period turned out to be one to never forget, and that was not only because of the new publication in our favorite journal.

In between writing sessions, I visited the residential area around the magnificent Parque das Nações and saw that new beautiful apartments were going to be built there. I had long dreamt of having a place of my own in this beautiful city, but I had never made this idea very concrete (the high prices had something to do with it). But after much discussion with my wife, we decided this time not to hesitate anymore, and we bought an apartment that we had never seen before (because construction had just begun). Before traveling back to Belgium, I shared the good news with José, who enthusiastically remarked that we would probably be able to develop many more research ideas now in a much more efficient way.

Just a few hours before I headed to the airport, as I bid goodbye to José, we suddenly got an idea for a new study that we promised each other we would work on next summer in the new apartment.

No wonder that this new idea would become our most beautiful one.

8

Can We Make the Projects Easier? (Part 2)

Note: This chapter is based on the article "*Reducing the feasible solution space of resource-constrained project instances*", published in *Computers and Operations Research*.

The two years following the previous study were years in which all sorts of things happened. We wrote our most challenging paper, which ultimately, from our perspective, would become our very best paper. It also became the paper that was accepted by reviewers with minimal changes. Despite the work taking somewhat more than a year, once submitted, the revision process lasted less than half a year and required only one round of revisions to get the paper published. I had not experienced something like that often, as reviewers often have a whole list of comments, often very justified, that take much more than just a few months of work. Additionally, I had set up my new apartment so that my wife and I could work more frequently from this beautiful city. We were both ready for our work life 2.0, and we planned multiple trips between Belgium and Portugal on a regular basis to divide our work life between these two places. And of course, I saw José more than ever during that time, and despite the fact that this study has become the very last study with José that will be discussed in this book, I am convinced that there is still a beautiful research future ahead of us, with many other additional studies (but that is for later, in another book).

I describe this paper as our most difficult for several reasons. One of them is that it took quite a long time for us to decide how to approach the new research. But the fact that I had too much on my plate trying to balance both my work in Belgium and the research in Lisbon while also dealing with a move from the north to the south of Europe likely played a part as well. In short, here is how our new study unfolded.

Summer 2022 I was back in Lisbon to work on our new research idea. I had bought my own apartment, but for the time being, I was still staying elsewhere. The construction phase was finished, now awaiting approval from the city council, which can take quite some time, they told me, in Lisbon. I had rented an apartment next to Parque Infantil da Praça da Alegria, meeting with José regularly to work out our new idea. Somehow, our ideas were all over the place, and despite considering this one of our strengths, this time we did not really reach a specific eureka moment. That is the nature of research, of course, and maybe it was because I was too busy daydreaming about what it would be like to have my own place in Lisbon. Dreaming about no more hassle with bad Internet or poorly lit kitchens in rented places, but my own space specifically designed to nurture research ideas. I was clearly looking forward to my new future residence where I would truly live and work, but for the time being, all we did that summer was explore new ideas that led to little progress.

Winter 2022 I finally had access to my brand new apartment in Parque das Nações. During the first few weeks of my stay, little research was done, as I had to furnish it to optimize the combination of living and working. But once this was sorted, I had time to discuss with José again. After some time, we realized that our ideas from the previous summer were not so bad after all. It had been good to let them rest for a while (during the fall), but suddenly we saw a promising direction. It was time to get to the real work now!

8 Can We Make the Projects Easier? (Part 2)

Spring 2023 I was back in Lisbon for about three weeks and had by now grown accustomed to the luxury of having my own place. We were both very excited about our idea, and even though the initial experiments on the supercomputer had been somewhat disappointing, we had learned a lot from them. During the spring stay, we had finally made up our minds and decided to expand on the "*solution equivalency*" concept from Chap. 6 and make the projects really easier this time.[1] Easier said than done, but this time, failure was no option.

Summer 2023 Our summer of 2023 was like Bryan Adams' summer of 69. Simply unforgettable. It was the first summer in Lisbon where I could enjoy the beautiful weather and walk to work from my own place, which I had meanwhile fallen in love with. Every day became one where our ideas became more and more promising, and each experiment led to more and better insights. That summer was perhaps the most intensive summer we had ever experienced together (from a research perspective, at least), and by the end of August, just as I had to return to Belgium, we had completed our study and submitted it to our favorite journal. As I mentioned earlier, the revision took just five months, and after only one round of revisions, the paper was ultimately accepted for publication in January 2024 in *Computers and Operations Research*.

This final chapter of the first part (Side A) of this book summarizes our last study (for now, at least) in our quest for more and better project data to better understand the well-known resource-constrained project scheduling problem. Do not hesitate, I am sure it is worth reading.

8.1 Solution Equivalency (*Extended*)

The solution equivalence concept from Chap. 6 continued to haunt us in our thoughts.

---

[1] The attentive reader may recall that this was also a goal in Chap. 6's study, but due to changes in resources, the projects did not get easier, but rather more predictable.

The idea of modifying the resource parameters of a project without altering the network seemed very original to us at that time. We were very pleased that we could demonstrate in this study that the existing resource parameters were not very reliable. However, it remained somewhat frustrating that this did not enable us to solve any project better or faster. These changes only altered the values of the resource indicators, allowing us to better discern phase transitions, but they did not reduce the total search space at all. Therefore, it was not surprising that few improvements could be found for these new projects, as each procedure searched within an equally large space of possible schedules.

But what if we were to search for changes in projects that would actually reduce the space of schedules? Without removing the optimal schedules of course! Would that make these projects easier?

It seems like a logical continuation of the solution equivalence concept from our previous study, but it took several months before we could translate this into a good research plan. We initially attempted to adapt the transformation procedure from Chap. 6, as beautifully presented in Fig. 6.2, but nothing came close to our objective. Since this procedure did not remove any schedule from the project being transformed, none of the adjustments were useful for our new idea.

Until we considered modifying the network of the project as well.

It is well known that projects with a more serial network structure are often easier to solve than parallel projects because resource conflicts occur less frequently. Therefore, the idea of modifying the networks of the projects to make them more serial, by adding extra precedence relations between activities, could certainly be a step in the right direction. However, as elegant and simple as the idea may be, it was also essential that the optimal schedule of the modified project remained the same as that for the original project. And therein lay the challenge of our latest research project.

The ultimate breakthrough that shaped our new challenge came when we introduced two new extensions of the solution equivalence concept. We called them "optimal equivalency" and "subset optimal equivalency" as depicted in Fig. 8.1. The basic idea of these extensions is roughly the same as the original solution equivalence concept, with the difference that

8 Can We Make the Projects Easier? (Part 2)

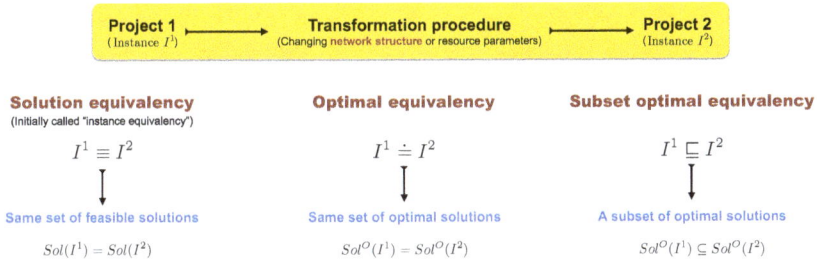

Fig. 8.1 Three times equivalency (changing network and resources)

the requirement for all schedules of both projects to be the same was reduced to only a subset of these schedules. The general idea is that a project instance I^1 can be modified into an instance I^2 in various ways, but as long as the optimal schedule of both projects remains unchanged, the project with the smallest search space could be considered the easiest project. This reduction of the search space while maintaining optimality forms the core and starting point of these two new equivalence concepts, and both differ only in one small detail.

To clearly distinguish the three nearly similar definitions from the figure, they are briefly explained below:

- **Solution equivalency**: Two project instances are considered solution equivalent if they share the exact same set of feasible schedules.
- **Optimal equivalency**: Project instances are optimal equivalent if they share the exact same set of optimal schedules, though they may have different sets of feasible schedules.
- **Subset optimal equivalency**: Project instances are subset equivalent if they have at least one optimal schedule in common, regardless of whether they share all optimal schedules or not.

The differences really lie in the details, but details often make a significant contribution to academic research. The idea of these extensions is very simple, but their implementation required the necessary caution to maintain optimality and the power of—once again—a supercomputer.

8.2 Cover Sets

To structure the search for new projects with at least one common optimal solution, we decided to partition the search space of projects into smaller parts. We did this using the so-called cover sets and five new propositions, which we eventually integrated into a new transformation procedure. It felt like everything from our previous studies was coming together, and we wondered how a new apartment in the most beautiful city in the world could have such an influence on our creativity. We did not know the answer, but we relished it nonetheless.

However, before the joy came, we faced several challenges that are briefly summarized below.

We started our research with the general idea that the search space of schedules for project I^1 can be partitioned into smaller parts by defining a cover set T for this project. A cover set T consists of a collection of $|T|$ project instances that together form the original instance I^1, as shown in Fig. 8.2. The solution space of a project instance I^1 is represented by $Sol(I^1)$ and contains all feasible schedules of that project. Therefore, the definition of a cover set implies that the search space of all possible schedules for instance I^1 is equal to the sum of the search spaces of the different project instances $I^{t_1}, I^{t_2}, \ldots, I^{t_{|T|}}$ from the cover set T. Each project instance from the cover set thus has its own smaller search space of solutions that together form the entire search space of the original instance I^1.

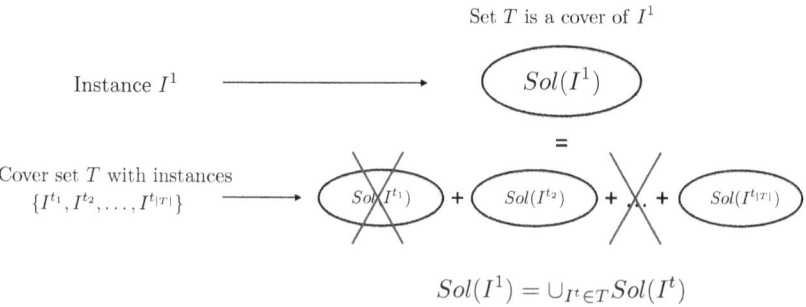

Fig. 8.2 Cover set

So, the cover set definition simply entails nothing more than a partitioning of the large search space into smaller parts. And if such a cover set can be found for a project instance I^1, then the goal is to remove as many instances from this set as possible, hence the crosses for a number of sub-instances in the figure, without losing the optimal schedule. Only in this way can the search space of the instance I^1 be reduced, because it can then be replaced by the remaining non-removed instances that still contain at least one of the optimal schedules of I^1.

It is once again a fairly simple idea, but the challenge remains, as it now comes down to constructing cover sets for projects, which turned out to be not as straightforward as initially thought. And yet, I believe that this perspective on our new research was the breakthrough, with the rest mainly consisting of a lot of work and even more computer experiments.

We decided to construct various cover sets for project instances by adding precedence relations to the original instance I^1 using five new propositions. Each of these propositions defines the cover set in a different way, but each time they guarantee that the totality of the sub-instances from the set is equal to the original instance. The basic idea of the first proposition is easy to explain, which I will do below with a simple example, but each extension to the four subsequent propositions brings additional complexity that I prefer to reserve for the readers of the publication of this study. Without delving into unnecessary technical details, an overview of these propositions is provided in Fig. 8.3 and illustrated, as promised, in the simple example below.

Imagine a project in which two activities, denoted as "a" and "b," can be scheduled in parallel according to the network, but the limited availability of resources prevents this. This means that scheduling these activities in parallel may be feasible according to the network structure, but it is impossible in reality because it would lead to a resource conflict. These activities are "*precedence compatible*" but not "*resource compatible*",[2] and since the example involves only two activities, these activities form a so-called resource-incompatible pair. If such a pair of activities can

---

[2] The attentive reader may recall that precedence-compatible and resource-compatible activities were previously used in the study of Chap. 6.

	Propositions (Defining incompatible activities)			Cover set (Number of sub-instances using extra precedence relations)	
ID	Name	Abbreviation	Activities	# sub-instances	Precedence relations
1	Pairs	RIP	(a,b)	2	{a→b} and {b→a}
2	Two pairs	2RIP	(a,b,c,d)	4	{(a→b),(c→d)}, {(a→b),(d→c)}, {(b→a),(c→d)} and {(b→a),(d→c)}
3	Clusters	RIC	(a,b,c)	6	{a→b→c}, {a→c→b}, {b→a→c}, {b→c→a}, {c→a→b} and {c→b→a}
4	Triplets	RIT	(a,b,c)	6	{a→b}, {a→c}, {b→a}, {b→c}, {c→a} and (c→b)
5	Quadruples	RIQ	(a,b,c,d)	12	{a→b}, {a→c}, {a→d}, {b→a}, {b→c}, {b→d}, {c→a}, {c→b}, {c→d}, {d→a}, {d→b} and (d→c)

Fig. 8.3 Creating cover sets for incompatible activity sets

never be scheduled together, then there are only two possible solutions to resolve this resource conflict. Either activity "a" is scheduled before activity "b" (a > b) or after (b > a), as there is simply no other possibility, and this forms the cover set of instance I^1. By simply adding one extra precedence relation from "a" to "b," the first sub-instance of the cover set is created, while the reverse relationship is added to create the second one. The solution spaces of both sub-instances together are equal to the solution space of the original instance I^1, and thus the cover set of instance I^1 consists of only two sub-instances, each comprising two different networks.

Simple as pie.

When this simple concept is extended to three or more activities, the same approach can be applied, with the difference that there are then more than two possibilities to resolve these resource conflicts, resulting in a cover set with more than two sub-instances. Notwithstanding these extensions to resource-incompatible triples (three activities) and quadruples (4), they still explore how resource conflicts for these activities can be resolved by adding precedence relations in the project network, leading to a number of sub-instances as depicted in the aforementioned figure. Same idea, just a bit more complex.

It is worth mentioning that we had to make a choice here, as after four activities, we could have continued to resource-incompatible quintuples with five and even to sixtuples with six activities, but the further we went, the less relevant these extensions became. As shown in the figure, with resource-incompatible quadruples, a total of 12 sub-instances are included in the cover set, and this number would only increase when the number of activities was further increased. Therefore, we decided to limit the number of activities to a maximum of four, and we made two additional extensions where we considered two pairs of resource-incompatible activities simultaneously (2RIP) as well as clusters of pairs from a total of three activities (RIC), but we did not go any further.

With these five propositions, we were able to build enough different cover sets for each project instance I^1 from which we could make a small selection without losing the optimal schedule. And we hoped that by doing so, we could significantly reduce the search space for each project, making these projects easier to solve.

8.3 Reducing the Search Space

With the different cover sets for a project instance I^1, a choice had to be made as to which sub-instances to retain and which to discard from each set, and this choice was made by a new transformation procedure that makes use of four known and three new theorems. The transformation procedure implements various changes to existing projects by adding precedence relations based on the propositions to construct the cover sets. It then selects only one or a few sub-instances of the cover set to reduce the search space of the instance. This process is sequentially repeated until no further changes can be found, ultimately resulting in a new modified project with a much smaller search space. A visual flow of this procedure is provided in Fig. 8.4.

Such an approach was, of course, not new for us as we had already modified projects in Chap. 6. Moreover, our new transformation procedure also utilizes the four theorems of the transformation procedure of Fig. 6.2, as can be seen in the top block mentioning theorems 1 to 4, explaining why the suffix "Part 2" was added to the current figure. There is, of course, nothing wrong with reusing parts that have been used before and are functioning well, but the real innovation lies, of course, in

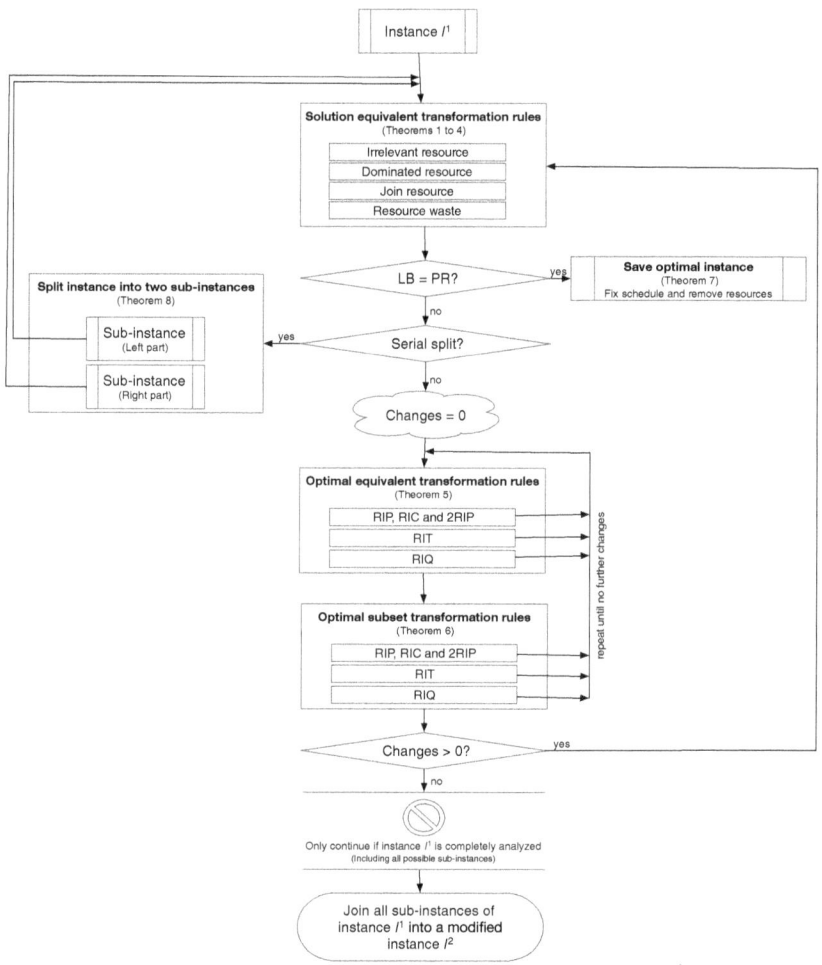

Fig. 8.4 Transforming projects (Part 2)

the parts that come after this initial step, which make use of three new theorems and the equivalence definitions discussed earlier.

The first four theorems, in fact, did nothing but modify the resource parameters and did not yet make use of the cover sets. It is only after these modifications that the cover sets for the projects are established, used by the three additional theorems, introducing further changes that alter the network structure through additional precedence relations. These three new theorems formed the core of this updated transformation procedure and are briefly discussed hereafter.

Theorem 5 The so-called optimal equivalence theorem defines the conditions to retain the sub-instance of the cover set that contains all optimal solutions of the original instance I^1. Specifically, the theorem checks if there exists a sub-instance for which a lower bound on the makespan is smaller than (or equal to) a known upper bound (representing the duration of a known schedule), while all other sub-instances have a larger lower bound (with a value greater than the makespan of the known schedule). If this is the case, it means that the optimal solution lies in the feasible space of that one instance, because all other instances only contain schedules with longer durations. In this case, that one sub-instance will be retained, and all the others will be discarded, leading to a significant reduction in the search space without losing optimality.

Theorem 6 The so-called optimal subset equivalence theorem does roughly the same, although the conditions are more stringent, and selecting one or more sub-instances now ensures the retention of at least one optimal solution. This means that some sub-instances may be discarded, even if they contain an optimal schedule too. The conditions can vary considerably and are not explained in detail here because that would lead to endless details, but the general idea can be described as follows:

Assume that the cover set consists of a number of sub-instances that can be put into two classes T_1 and T_2. If for T_1 instances from the cover set T a feasible solution can be found with a duration equal to the best-known duration available in the literature, and the other T_2 instances have a lower bound makespan value that is equal to or greater than this known duration, then at least one optimal solution is available in the sub-instances of T_1. And even though the instances from T_2 might contain some optimal solutions too, they will not be retained. It is important to note here that the conditions of Theorem 6 can only be applied if a good feasible schedule is known from the literature. Sometimes this is indeed known, but even if it is not, such schedule can always be generated with existing scheduling algorithms. It may be the optimal schedule, or not, and usually, we do not know, but that is not very important because the conditions of Theorem 6 always guarantee that at least one optimal solution remains retained.

These two new theorems fully utilize the extended equivalence definitions discussed earlier. When the conditions of these theorems are met, a portion of the cover set is cut away, but the remaining instances have extra precedence relations, which naturally change their network structure compared to the original instance. Subsequently, this process is iteratively repeated to make further changes to the cover sets of these new project instances until no further changes can be made. The ultimate result is a much smaller solution space, with no loss in the optimal solution, but it is computationally very demanding. Due to the enormous number of possible changes, this transformation can become an immensely intensive process. Moreover, as the cover sets for each project can be generated in five different ways using the five propositions, and as the project grows in size, the number of possibilities becomes so vast that even a supercomputer would struggle with it.

But it was all worth it, as a smaller solution space implies an easier project, which was our ultimate goal.

Theorem 7 And the hope that many of our modified projects could become much easier led to the implementation of our seventh theorem. This theorem checks whether there exists a lower bound value for the project makespan that is equal to the duration of an existing schedule. If that is the case, then the project is already solved anyway, which is the ultimate proof that it has become an easy project.[3] The lower bound is nothing more than an underestimation of the scheduled project duration, whereas the duration of a real schedule can easily be found by applying priority rules. These priority rules can build a feasible schedule in a very easy and fast manner, although its quality is often not very high. Hence, the likelihood that the lower bound is equal to the makespan obtained by such a quick schedule is not very high, and Theorem 7 does not seem very relevant.

---

[3] Recall that the project instances for which a lower bound is equal to an upper bound are called *closed* instances, and it is thus proven that the schedule is the best possible schedule.

Yet, that is only an illusion.

That is because the transformation procedure does not always operate on entire projects but also on parts of projects. As a matter of fact, we implemented an additional rule in our procedure that ensures splitting a project into two halves, and the procedure continued to work on each half separately. Indeed, a project with two phases executed sequentially can just as well be considered as two smaller projects, and then Theorems 5 and 6 can also be further applied to these separate projects. Thus, despite the fact that priority rules do not easily lead to good schedules, this chance could be significantly increased by working on parts of projects, ultimately enhancing the practical utility of Theorem 7.

And all of this led us to the new transformation procedure that José and I are still very proud of.

I may have overlooked some technical details, but the summary above describes the core elements of our new transformation procedure. If it seems like all of this is an easy task for a powerful supercomputer, then for us, it sometimes felt like a nightmare to implement and test this procedure. There were so many possibilities for making changes, that it seemed like an endless search where too many choices had to be made. But we made many, and once the right choices were made, the procedure could be run without too many problems. Our ambition resurfaced, and we decided to test it on 11,183 of the project instances available to us.

Without a supercomputer, that would have taken years.

We accomplished it in two Lisbon summers.

8.4 Four Times More Data

The modifications made to the 11,183 project instances resulted in 9421 new instances that shared at least one optimal schedule with the original projects. This means that for 85% of the projects, new modified projects were found, while for the remaining 15%, our transformation procedure was not successful. Out of our 11,183 instances, 9722 already had an optimal solution known in the literature (closed instances), making the changes less relevant. However, for the remaining 1461 instances where optimality had not been proven yet (open instances), 370 (25%) could be modified using our new transformation procedure. We hoped that due to their smaller search space, these modified open instances would become easier to solve too.

Our hope gradually turned into reality as we progressed through the study.

We now had four versions for each project, leading to a total of four times $11{,}183 = 44{,}732$ projects (although 15% remained unchanged, being identical to their original version). In a previous chapter, we already created the "lowRU" and "highRD" versions for each project, which only modified the resource indicators but retained the same search space. Additionally, through our new study, we added a fourth version of each project and called it the "keepOPT" version. Our first breakthrough came when we observed that 7482 of the 11,183 project instances (67%) could be optimally solved after the modifications without the use of any scheduling algorithm, purely because the lower bound and upper bound values became equal, proving optimality with Theorem 7. This was a significant achievement, particularly for the 18 open instances that had never been optimally solved before. José and I celebrated with a good beer for these noteworthy results.

Eureka!

Finally, projects were more easily solvable.

But our hunger for further progress was not satisfied yet. We wanted more. For the remaining 3701 (11,183–7482) projects, optimality could not be demonstrated through the modifications alone. Yet, we believed that these projects might have become easier to solve too, and we were determined to prove it. It should be possible to find good or even optimal schedules for these projects quickly as well. Otherwise, our initial breakthrough would remain just that—a single moment of success.

And as they say, all good things come in threes, so we continued our quest for further discoveries.

8.5 Easier Projects

We decided to demonstrate convincingly that the 3701 remaining modified projects could also be solved more easily than their original versions in two completely different ways. I suppose we were keen on impressing the readers of our new publication with our results to fully convince them to use the "keepOPT" versions of the project databases from now on.

Method 1. Analysis of the Project Indicators The first method is one that we previously used in Chap. 6, where we examined whether the network and resource indicators of the modified instances ended up in the easy or complex zones. Since the project modifications of the transformation procedure alter both the network and resource indicators, it is interesting to see in which direction their values were evolving. From the phase transitions research, certain intervals are known for these indicators where projects are difficult, and other intervals where they are much easier. We therefore want to determine whether the modifications have brought the indicator values of the modified projects into the easy or difficult zones. The results of this analysis can be found in Fig. 8.5 for both the network indicators (top) and the resource indicators (bottom).

We did not examine all the indicators, but rather selected a few known to be quite reliable. For the network structure, we focused on both the order strength and the serial/parallel indicator, as they both measure how closely a project resembles a complete serial or parallel network. The figure illustrates that both indicators showed higher values for the modified project instances (y-axis) compared to the original instances (x-axis), clearly indicating that the modifications resulted in easier instances. This was not unexpected because we generated the instances of the cover sets by adding extra precedence relationships each time, leading to more serial networks, which are known to be easier. However, this positive outcome was further confirmed by the resource indicators, which also shifted toward the easier direction. As readers may recall from previous chapters, we used the resource strength, the resource factor, and the resource constrainedness as the most well-known resource indicators. However, José and I, along with co-author Rob Van Eynde, conducted research proposing two new resource indicators, called FS21 and FS31 (with FS representing feasible sets), which could better predict project complexity and thus were more reliable. These indicators showed that lower values represented easier projects. Consequently, we observed that for these indicators, the modified instances were easier than the original

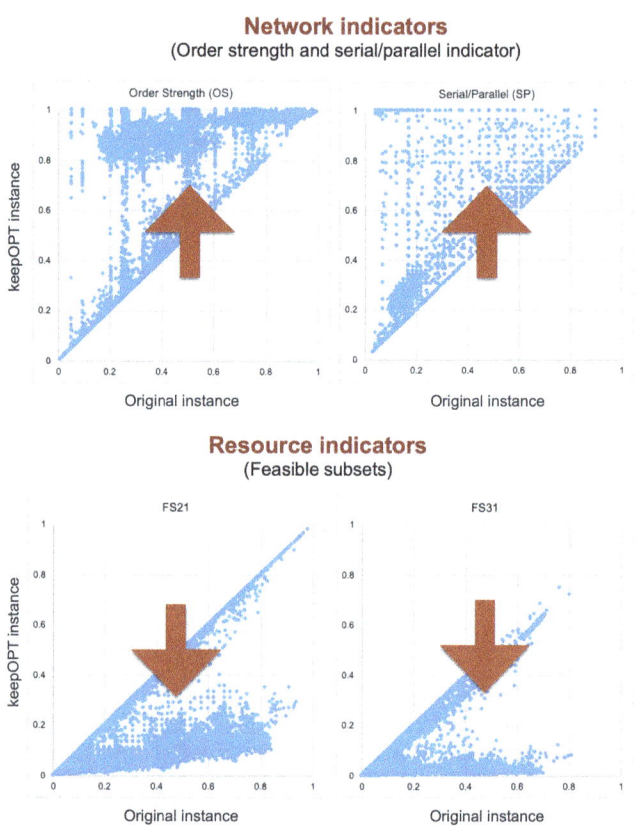

Fig. 8.5 Easier projects (predicted by indicators)

instances, as all their values decreased. Since all indicators pointed toward values in the easier zones, this marked a second significant breakthrough and served as proof that we had indeed made our projects much easier. It was time to celebrate with an additional Superbock beer on a sunny terrace in Lisbon.

Eureka again!

Nothing but good news so far, but the indicators obviously do not show the whole picture. They can somewhat predict whether a project is more likely to be easy or difficult, but the real proof that the modified projects have become easier can only be provided by actually tackling the challenge of finding very good schedules for them. And for that, we obviously need scheduling algorithms and a supercomputer.

But fortunately, we had all of this at our disposal.

Method 2. Analysis of the Solutions by Scheduling Algorithms From our previous studies, we had the 48 branch-and-bound procedures from Chap. 4 that could obtain optimal solutions, as well as the three meta-heuristics from Chap. 7 that had already proven to generate very good, though not necessarily optimal, solutions. With these two classes of algorithms, we wanted to demonstrate that the modified projects could be solved much better and faster than the original projects, and we did this by initiating another extensive computer experiment. Needless to say, the supercomputer played a crucial role in this endeavor. It has always been like that in our research over the past few years.

The results were stunning.

The results confirmed that the modified projects were indeed easier to solve, for both the exact and heuristic algorithms. For the exact branch-and-bound algorithms, we could find significantly more optimal solutions for the modified projects compared to the original projects, as shown in the left graph of Fig. 8.6. This figure only displays part of the results by comparing the three branching schemes discussed earlier in Chap. 4 (activity-based (ACT), serial (SER) or parallel (PAR) branching schemes). The parallel branching scheme still performed significantly

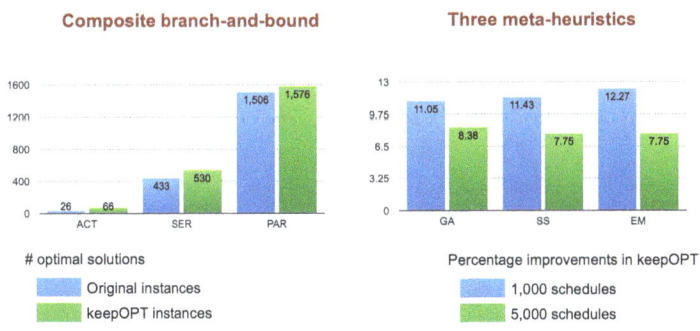

Fig. 8.6 Easier projects (proven by algorithms)

better than the other two, as we had seen in previous studies. However, the number of optimal solutions found within 1 minute was much higher for the keepOPT instances, clearly demonstrating that they became much easier.

For the meta-heuristics, the news was also unequivocally positive, as suggested in the right graph of the figure. This graph shows the percentage decrease in project makespan of the modified projects compared to the original projects when these three procedures are terminated after 1000 and 5000 schedules. The graph only shows positive values, indicating that the keepOPT project schedules have much lower makespans than the original projects after an identical search, and the improvements are significant. For 1000 schedules, which is only a small fraction of all possible schedules, improvements of more than 12% could be found, which is quite substantial. For the somewhat more intensive search of 5000 schedules, the differences were smaller, but that is all relative because they all still amount to more than 7%.

Nothing but good news, and I am not even done yet.

When we used both algorithms again, but this time in a very intensive manner with long computation times, we found improvements that had never been discovered before. We found 16 improved schedules in our modified projects, and we could even prove for 64 projects that the solution contained the optimal schedule, transforming these open instances into closed instances. Additionally, we also found many improvements in the lower bounds for 110 projects, and it goes without further explanation that both José and I, drenched in sweat from our hard efforts, proudly look back to this day on such an intense research project.

This last study finally provided the ultimate proof that the manipulations of projects can indeed be fruitful. We had managed to simplify our projects!
We had found better schedules!
Third time eureka!
We had finally achieved our main goal.

8.6 Mission Accomplished

Indeed, we had finally achieved what we set out to accomplish. For years on end, we had investigated data for projects, modifying our projects, refining them, cleaning them up, even manipulating them, all with the same goal in mind: We wanted not only to better understand the complexity of the RCPSP but also to solve it better than ever before. It took a while, but eventually, we found a way to make the projects easier for this challenging problem. And we eventually also found some better solutions for them. We have succeeded in our endeavor, at least partly. Partly indeed, because the problem still holds many secrets. Research is never a finite story.

This last study became our pinnacle, as its publication marked our fastest ever. I know I wrote about this before, but the incredible speed at which this paper was accepted had never happened to us before. With a turnaround time of just a few months and after only a single revision (indeed, just one), the paper was published in early 2024 in, as you might guess, *Computers and Operations Research*. I was back in Belgium when I received the good news and immediately informed José about it. I also promised to come back to Lisbon soon.

He was naturally as enthusiastic as I was, and we agreed that our next meeting would be one to cherish the results of our hard work. No more research, just enjoying the aftermath of our intense and rewarding collaboration. That was the plan! And right after making that promise, I sent this book to my publisher.

They say that all is well that ends well. It marks the conclusion of a beautiful story in pursuit of data for projects, set in the two most beautiful cities in the world.

So that was it. For now, at least. For the book, anyway.

But we are only in our early 50s.
Too worn out to get into cycling.
Too old to start our own progressive rock band.
But teenagers when it comes to a research career.

And so we know . . .
. . . the best is yet to come.

Part II

Side B. The Relevance of Project Data (Featuring Other Researchers)

The quest for improvements through the continuous manipulation of artificial project data, which I described in the previous part of this book, remains one of the main objectives of my research, conducted in collaboration with José Coelho. It is this never-ending quest for insights and solutions to the well-known resource-constrained project scheduling problem that continually prompts us to consider alternative approaches. These results are extensively discussed at workshops and conferences. Among academics, you would find our discussions quite intriguing and intense at times, and sometimes even a bit otherworldly.

However, despite this pursuit, I am certainly not oblivious to practical concerns. Given my extensive teaching commitments at business schools and interactions with companies, I have deliberately sought out practical applications during my career, and this search for practical relevance also shows up in my pursuit of understanding project data. That is why in this second part, I will explore the project data world more deeply and go beyond just looking at the RCPSP.

In Chap. 9, I will extent the narrow view on project scheduling to project predictions under uncertainty. As a matter of fact, it is often crucial in practice to predict the total duration and/or costs of projects, even without having a schedule, and this can be best done using real project data. Although I solely used artificial project data in the previous book part, I also made a significant effort in my research group to collect empirical projects, which will be introduced in this second part. It is not difficult to understand that with real projects, predictions are more accurate than with artificial projects simply because they involve real data, real uncertainty, and real risk. However, there is a difficulty hidden within this approach. These predictions rely on statistical techniques, which require stochastic distributions for project activity durations, the values of which are often not known in practice. That is why I propose a so-called calibration method to search for the correct parameter values of stochastic data points using these real projects. This study is not undertaken with José, but with Jeroen Colin and Jordy Batselier, who were part of my fantastic team in the Operations Research & Scheduling group at the time of the research.

In the final Chap. 10, I conclude this book by providing an overview of both the artificial project data from the previous book part and the empirical data from this part. I also demonstrate in this chapter that the RCPSP, the challenging problem I have been working on with José for so many years, is just a well-defined problem with many possible variants. This chapter takes a journey toward extensions such as people skills, alternative technologies, and project portfolios. You will also get to know other members of my research group, such as Tom Servranckx, Jakob Snauwaert, and Rob Van Eynde, all fantastic people without whom I could never have written this chapter.

And oh, before I forget, José appears several times in this very last book chapter.

How could it be otherwise?

9

How to Make the Project Data Practically Relevant?

Note: This chapter is based on three studies performed by different members of the Operations Research & Scheduling group. First and foremost, the original calibration procedure was validated using empirical project data in the study titled "*Empirical perspective on activity durations for project management simulation studies,*" published in *Journal of Construction Engineering and Management*. Next, the calibration procedure was extended to human partitioning in the study "*Fitting activity distributions using human partitioning and statistical calibration,*" published in *Computers and Industrial Engineering*. Last but not least, the procedure was extended to a powerful statistical partitioning in the study "*A statistical method for estimating activity uncertainty parameters to improve project forecasting,*" published in *Entropy*.

In the previous part of this book, the focus was on using artificial projects to gain a better understanding of the complexity of the resource-constrained project scheduling problem. Through the generation and modification of projects, we searched for both complex and easy projects with scarce resources. Using our supercomputer, we conducted tests that would typically take years if we did not have access to such a powerful machine. Our journey of over 20 years ultimately led to insights into this challenging problem and, ultimately, to better solutions. Never did we question during this research journey whether we should also use real project data. The majority of our research was conducted using artificial project data, and that kept us more than occupied.

But the ultimate goal of academic research is, of course, to generate insights that are extracted from the ivory tower and prove useful for the practical management of projects. This ambition can only be achieved if academics also venture beyond their safe environments to seek real data of real projects, not just artificial ones. One might begin to wonder why we never did this in the previous part of this book. Perhaps this lack of realism diminished the value of our previous research considerably.

But I do not think so.

Sometimes, artificial project data proves to be much more suitable for research than empirical projects. But that is not always the case.

In the research on scheduling projects with limited resources, there is little need for real projects. The scheduling algorithms do not require much data anyway. With a list of activities and precedence relations forming the project network, along with a set of resources with limited availability and demand from the activities, most algorithms can get to work. And whether these data come from real projects or artificial projects, it does not matter much. Besides, with artificial projects, we can control these data much better and modify them in various ways to better search for the essence of project complexity. That is what we did in the previous part of this book, by the way.

However, managing projects requires more than just creating a good project schedule.

A project schedule is merely the starting point in a much more significant phase of the life of projects. Since this schedule can be distorted by various disruptions during project execution, it serves only as a reference point to manage these disruptions as effectively as possible. Therefore, academic research also pays significant attention to analyzing potential issues with such a schedule and how such disruptions can be easily detected to take timely actions. This risk analysis of the project schedule is known in research as *"schedule risk analysis,"* typically investigated by simulating artificial delays in activities to identify which activities are most sensitive to such disruptions. In addition to analyzing potential disruptions, research also focuses on timely detection of these disruptions in various forms, all classified under the umbrella of "*project control*" research. Often, techniques like earned value management are used to measure project progression in a relatively straightforward manner. When this system detects significant deviations from the project schedule, it serves as a warning signal that it is time to take action to get the project back on track.

A project schedule is indeed important, but it is only the beginning.

The integration of project scheduling, risk analysis, and project control is known in the literature as "*dynamic scheduling*," and throughout most of my research career, I have tried to combine these three components. In the majority of my previous books, for which I provide an overview at the end of this book, the integration of these three components is central. Only in the previous part of this book, did I focus solely on the scheduling dimension.

Since I have already discussed these topics in several previous books, I will not revisit the details of *schedule risk analysis* and *project control* again. However, what is important to know is that most research in these two domains uses probability distributions on activity durations to simulate the potential execution of a project. These distributions are then used in Monte Carlo simulations to mimic the possible project executions, generating a wealth of data for researchers to analyze and derive new insights from. These probability distributions reflect the uncertainty in activity durations and are often chosen quite arbitrarily since the exact uncertainty is naturally not known. Researchers, therefore, utilize various distributions, ranging from symmetric distributions like the normal distribution to skewed distributions like the beta or the lognormal distribution, without necessarily knowing if they accurately reflect reality. The choice of parameter values for these distributions exacerbates this issue since these values are also typically unknown. As a result, the mean values and standard deviations of the distributions are also chosen quite arbitrarily, and as long as they are varied sufficiently, they might accurately mimic reality.

But is this truly the case?

This chapter tries to provide an answer to that question.

A lengthy argument is unnecessary to illustrate that in risk and control research, the most effective approach is to examine real-world scenarios, involving projects with genuine delays and disruptions. The value of empirical project data lies in the undeniable occurrence of delays, which are not merely random events as portrayed in artificially selected distributions, but often have identifiable causes.

In the research studies of this chapter, we utilized real empirical project data to select activity duration distributions in a more realistic manner. By employing calibration techniques, we transformed historical project data into statistical distributions with predetermined parameter values, thereby enhancing our understanding of these distributions and improving the precision of our Monte Carlo simulations. Our focus expanded beyond artificial projects and scheduling algorithms, broadening our research perspective to make project data practically relevant.

We wanted to build a bridge between theory and practice.

9.1 Calibrating Data

The idea of data calibration is straightforward yet ingeniously devised. While I wish it were my own idea, credit must go to a study by Dan Trietsch and his co-authors published in the *European Journal of Operational Research* in 2012. In their work, they proposed a novel method to determine the distribution parameters of activity durations in projects based on only two inputs, as illustrated in Fig. 9.1.

Primarily, these calibration procedures rely on the planned durations of activities as an input, which are estimates provided by project managers during the project schedule design phase. In previous chapters of this book, we simply generated these numbers randomly. However, for real projects, thoughtful consideration must be given to these estimates. Additionally, alongside the planned activity durations, there must also be actual activity durations available, which, naturally, are only known for real projects once the activities are completed. The values of these two durations are likely to differ due to unknown risks, potentially resulting in delays or, perhaps, occasional instances of activities being completed

9 How to Make the Project Data Practically Relevant?

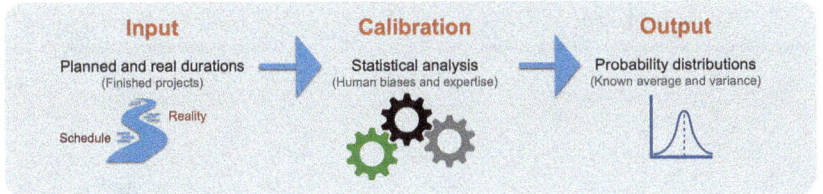

Fig. 9.1 Calibrating real project data

on time or even ahead of schedule. The source of the deviations between planned and actual activity durations does not matter much, as they arise from real observations and therefore reflect genuine disruptions in real projects. No more theory, just empirical observations where sufficient disruptions occurred.

Fortunately, in the real world, Mr. Murphy is omnipresent. Disruptions are more the norm than the exception.

The proposed calibration procedure analyzes the variances between planned and actual durations for each activity, aiming to estimate distributions through statistical methods. While similar analyses are common in literature, typically involving statistical fitting methods to find the best-fitting curve, the calibration method goes further by extending beyond classical approaches. This unique extension caught my attention and sparked my interest in its potential applications.

The unique concept of data calibration stems from recognizing that the data obtained on activity durations may not always be equally reliable. This is because both planned and actual durations are reported by project managers or their team members, and errors, whether conscious or unconscious, can occur, potentially compromising the data quality. The calibration method addresses the issue of human bias in the reported data by removing unreliable data points, aiming to achieve better parameter estimates of the distributions. The removal of unreliable data points from empirical projects appears to be entirely new in the literature, and I believe that such a data cleanup process is what the authors refer to as "*calibrating the data.*"

It is a statistical method that dares to question the reliability of those providing the data.

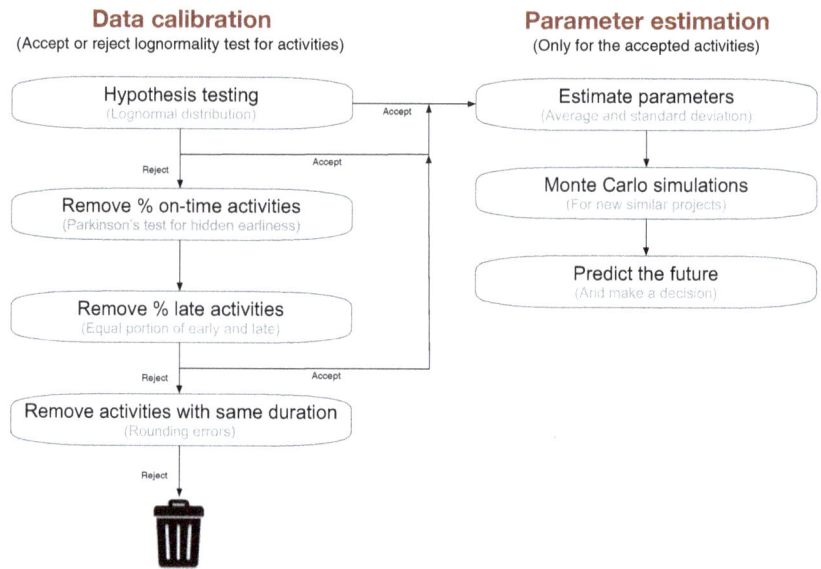

Fig. 9.2 Sequence of hypothesis testing

The method assumes that the activity durations follow a lognormal distribution, which is an acceptable assumption in the literature, and iteratively tests whether this is indeed the case. More specifically, this step-by-step method iteratively analyzes the activity durations, removes possible unreliable data points, and repeats this process until ultimately a remaining portion of the project activities is reliable for parameter estimation. The output of such estimation consists of average values and standard deviations for the lognormal distributions of the remaining activities, reflecting the uncertainty in the activity durations. Just as researchers previously made these estimates based on arbitrary choices, they are now obtained from real project data with genuine delays.

This iterative process is depicted in Fig. 9.2 and comprises a series of statistical tests to ascertain whether the lognormal distribution can indeed be assumed for certain activities, with the removal of those for which this assumption cannot be accepted. The entire process consists of four sequential phases, as indicated in the left part of the figure, which are briefly summarized in the following lines.

Step 1. Hypothesis Testing Initially, the assumption is tested for all activities of a project, with the null hypothesis being that the ratio of the actual activity duration (RD) to the planned duration (PD) follows a lognormal distribution. This ratio categorizes activities into three groups: early completed activities (RD/PD < 1), on-time activities (RD/PD = 1), and late activities (RD/PD > 1). This test is conducted using a classic normality test because if the ratio follows a lognormal distribution, then the logarithm of this ratio must follow a normal distribution too. Based on the obtained p-values, it is then determined whether this assumption can be accepted or not. It is classic statistics, up to this point.

Step 2. Remove on-time Activities If the lognormality assumption cannot be accepted for all project activities, the calibration method seeks activities where the duration reporting was not reliable and removes them so that the lognormal test can be performed again on the remaining activities. The primary and most crucial reliability test is the Parkinson test for on-time activities. While some activities may indeed have been completed precisely on time, it is also common for them to have been finished earlier than planned, yet not reported as such by the team member. Reporting early completion would imply sudden availability of time, potentially leading to additional work for the team member without initial planning. Therefore, individuals often refrain from reporting this early completion and instead pretend that the activity was completed just on time, still satisfying the project manager. This phenomenon, known as the Parkinson effect, originates from a 1957 study in which Parkinson's law states that *"work expands to fill the time available for completion."*

Step 3. Remove Late Activities If the procedure in the previous step has removed a certain percentage of the on-time activities from the dataset, it may disrupt the reliability of the data. Because this removal is carried out assuming that these activities were actually early activities, it will disturb the original balance between the real early activities and the late activities. To restore this equilibrium, the same percentage of late activities is removed in a third step. This third step is essentially an intermediate step to ensure that the calibration procedure continues to operate realistically. However, after this step, the balance is restored, and the procedure will once again perform the lognormal test for the remaining activities. If accepted, the procedure stops with the confirmation that the remaining activities follow a lognormal distribution. If not, the procedure assumes that there are still unreliable data in the set and proceeds to the next step.

Step 4. Remove Rounding Errors In a final attempt to exclude unreliable data points, the procedure will conduct a second human bias test to determine if rounding errors were introduced for some activities. It is common for people to round the actual duration for convenience, resulting in many activities having the same duration. For instance, if the schedule was created with durations expressed in weeks, then an activity with a true duration of 4.5 days will often be reported as 1 week, which is actually an overestimation. To address this issue, activities with the same duration are also removed from the dataset, and the lognormal test is performed once again. For the very last time, when the lognormal test can be confirmed for the remaining activities, there is statistical evidence that the lognormal distribution is valid, and the procedure can proceed with estimating the parameters of this distribution. If the test was not successful, then the project data calibration failed, and the data for this specific project cannot be used for further analysis.

The estimation of parameter values for the remaining activities is not very difficult. Essentially, it involves dividing the remaining activities into clusters, where activities within a cluster receive the same values, while activities between clusters vary significantly. Since it has been statistically proven that the durations of these activities follow a lognormal distribution, it suffices to build these clusters for the average duration and the standard deviation of these durations. There are many techniques for dividing data points into clusters, but a classical K-means clustering is sufficient to do this accurately enough.

Once these values are known, it becomes interesting for researchers, but also practitioners can now predict their projects much better. For researchers, the added value lies in the fact that the Monte Carlo simulations they use to imitate project executions will now be based on real data instead of arbitrary choices, which improves realism. Of course, this does not mean that they cannot still test many variants, but at least they know where to start their research.

But anyone who thinks that Monte Carlo simulations are only interesting for academics is completely mistaken. Monte Carlo simulations are indeed very interesting for project managers, perhaps even necessary. As a matter of fact, Monte Carlo simulations offer project managers a powerful method for assessing and managing risk in their projects. By generating multiple scenarios based on probability distributions on calibrated data from their own past projects, these simulations can help them understand uncertainty for their future projects, make informed decisions, optimize resources, and effectively communicate with stakeholders. These simulations provide quantitative data to make better predictions and support decision-making, facilitate continuous improvement, and ultimately contribute to the success of projects by enabling proactive risk management and resource allocation.

And all of this is based on their own data, not on arbitrary estimates or well-informed guesses.

9.2 Empirical Validation

Impressed by the elegant simplicity of the calibration method, I embarked on an investigation to ascertain whether this approach could indeed yield reliable parameter estimations. Collaborating with Jeroen Colin, we investigated the data from 24 projects within our own empirical database presented earlier in Chap. 2. Despite the intricacies involved, we managed to determine parameter estimations for 12 of these projects, an outcome that we deemed moderate yet promising. Encouraged by our findings, we decided to document our research in a published study.

Initially, we were quite happy with our progress, anticipating that this method would pave the way for future simulation studies utilizing real data rather than artificial projects. However, our satisfaction was tempered by the realization that achieving a success rate of only 50% (12 out of 24 projects) fell short of our expectations. This realization prompted us to recognize that our initial publication on the calibration method should not be considered the final word on the matter.

As is often the case with breakthroughs in research, each advancement tends to raise more questions than it answers. Undeterred, we remained hopeful for improved results and continued our quest for a deeper understanding of the calibration process.

9.3 Human Partitioning

Despite the promising results yielded by the calibration method, we could not overlook the fact that half of the projects failed to find distribution parameters for their activities. It was perplexing to consider that so many activities had to be excluded due to the biases we previously identified. While we acknowledged the presence of hidden earliness and rounding errors within our data, we could not help but wonder if there was another underlying reason for the removal of so many activities from further analysis. Perhaps our approach of viewing the project as a uniform entity was flawed, overlooking the intricate composition of projects into smaller entities.

This realization led us to hypothesize that treating each project as a one-size-fits-all entity might hinder the success of the calibration process. We contemplated the idea of classifying project activities into groups, which we termed *partitions*, and conducting the calibration analysis on each group separately. Recognizing that many projects comprise multiple interconnected subparts, we understood the importance of not treating all project activities as one entity during the calibration procedure. Just as project managers break down their projects into subparts in the work breakdown structure, we aimed to do the same with the calibration method, albeit lacking a clear method of creating these partitions.

Our solution came from engaging with the project managers who provided us with the data. While their initial responses regarding partition criteria were disappointing, further exploration revealed that they had already implicitly partitioned their projects in various ways. We identified three key inputs that contained information for dividing activities into partitions: the baseline schedule, the work breakdown structure, and risk classes.

The *baseline schedule* provided insights into the duration and complexity of activities, leading us to distinguish between different classes of activities based on their scheduling characteristics. Some activities were scheduled for several days, while others were sometimes scheduled for several weeks or longer. We believed that these two categories of activities belonged to entirely different classes, and it would be somewhat futile to treat them as a single group with uniform parameters. The *work breakdown structure* facilitated a more structured division of the project scope into phases, with each work package considered as a separate partition. The division into work packages was particularly intriguing because it often followed a logical structure that delineated the project into distinct phases. On average, our data revealed eight work packages per project (though sometimes as many as 26), leading us to consider each work package as a separate partition, each containing a varying number of activities. Lastly, for projects where *risk classes* were defined by the project managers, we used them to create additional partitions, acknowledging the varying levels of risk associated with different groups of activities. These classes were primarily descriptive, categorizing risks as low, medium, or high without relying on much data or the distributions that we aimed to achieve. Nevertheless, they still indicated, to some extent, the presence of different risk groups within the project. Consequently, we decided to incorporate them into the partitioning process as well.

This so-called human partitioning phase is conducted prior to the calibration procedure, relying solely on human expertise without using any other method. It allowed us to divide projects into classes of activities, with each class undergoing calibration steps separately. While activities could still be removed from each class, the aim of this division was to preserve more activities and estimate a separate set of parameter values for the lognormal distribution of each accepted partition.

At the time of this research study, our empirical dataset had grown, and we conducted an analysis on 97 projects using this refined calibration approach. We were pleased to discover that 80% of the activities were retained after calibration, indicating that the lognormal distribution parameters could be accepted for the majority of the partitions. While our previous results were somewhat lacking, splitting activities into partitions emerged as a crucial step to enhance the calibration method. However, our quest for improvement was far from over. As researchers, we are driven by an insatiable appetite for further refinement and enhancement of our methods, never completely satisfied with our results.

We intuitively felt that there was just a little more in it.

9.4 Human and Statistical Power

In our third and final calibration study, we decided to not only leave the partitioning to the project manager who uses simple criteria such as baseline schedule durations, the work breakdown structure, or risk classes. Instead, we aimed to expand the statistical aspect to allow for "*automatic partitioning.*" This decision was inspired by the increasing focus on machine learning algorithms, which have the ability to discern patterns within data that may reveal hidden logic and enhance predictive accuracy beyond human capability.

Why leave it to a human being, if a machine can do it better?

But we did not go as far as using machine learning algorithms, although this may be considered for a future study. Instead, through easy statistical methods, we were able to discover patterns that would automatically perform this partitioning, clustering the activities and then initiating the calibration methods for each separate cluster. This automatic partitioning was based solely on the analysis of the data, searching for features to divide the activities into groups. Features that the project managers were not familiar with or did not see, but which statistical techniques were able to uncover.

And the results were stunning. This final extension further improved the accuracy of the calibration procedure, making it nearly perfect.

This time, we conducted our tests on 125 projects (our database kept growing) and were able to accept the lognormal distribution for almost all activities, reaching an accuracy well above 80% achieved by the human partitioning. Furthermore, we observed in the results that the

9 How to Make the Project Data Practically Relevant?

Parkinson effect was the primary human bias responsible for removing some activities, and the rounding effect did not occur as often. The outcome of this calibration process resulted in the creation of a relatively concise set of partitions for each project, typically ranging from one to six at maximum. Limiting the number of partitions is a crucial aspect in maintaining usability standards, ensuring that the system remains navigable and comprehensible for users. With too many partitions, each activity would be seen as a separate cluster, each with different parameter values (for averages and standard deviations). It would be very difficult to use the calibration results for new projects because then a choice for too many parameter values would have to be made in each separate cluster. However, with a small number of partitions, these choices are more limited, and the sensitivity to incorrect choices is also much smaller, thus improving accuracy.

This last extension is currently our final one, and we began using it in many of our studies, but we did not further expand upon it. What particularly led us to decide not to extend this automatic partitioning calibration method any further (for the time being) is that a combination of human and statistical partitioning yielded the best results, retaining 97% of the activities for which the lognormal distribution with known parameter values is accepted.

Our last calibration study has clearly demonstrated that extending the original calibration to include both human and statistical partitioning resembles an ideal collaboration between humans and machines,[1] but I truly believe that calibrating project data is still in its infancy and can open many new doors to narrowing the gap between theory and practice.

Perhaps we should indeed consider exploring an additional extension anyway.

---

[1] I referred to it in my book "*The Illusion of Control*" as the rider and the horse.

9.5 More Empirical Data

While artificial project data can be useful for certain research purposes, nothing can truly replace the richness and relevance of real empirical data. However, to harness the full power of real data, calibration is essential. By calibrating models and simulations using real empirical data, researchers can ensure that their findings are grounded in reality and accurately represent the complexities of the situations they are studying. This approach not only enhances the credibility and validity of their research, but also enables more insightful analysis and more informed decision-making in practical applications. So, while artificial data may have its uses, the calibration of models with real empirical data remains the gold standard for achieving meaningful results in project management research.

In Chap. 2, I introduced the empirical database that my research group had built up over the years. I mentioned that we initially started with 51 projects but eventually expanded it to a set of 181 projects, which I presented in my previous book "*The Illusion of Control*." However, just before sending this current book to my publisher, I reviewed my recent projects to see if there was anything useful among them. As a result, the database has grown once again, now consisting of 199 projects. You likely sense my pride in reaching this new milestone, and I hope to soon announce that our empirical set has grown again, simply because I strongly believe in the utility of this data, especially when linked to artificial projects, for academic research. The empirical projects come from various sectors, as shown in Fig. 9.3, and interested readers can download all the empirical as well as the artificial data from www.projectmanagement.ugent.be/research/data. It is free and conducive to academic research, so I see no reason why you should not take a look at our project data website.

Despite the proven usefulness of empirical project data for calibration, I want to reiterate my strong belief in the use of artificial projects for academic research. It may seem logical; otherwise I would not have dedicated the first eight chapters of this book to these artificial projects,

9 How to Make the Project Data Practically Relevant? 135

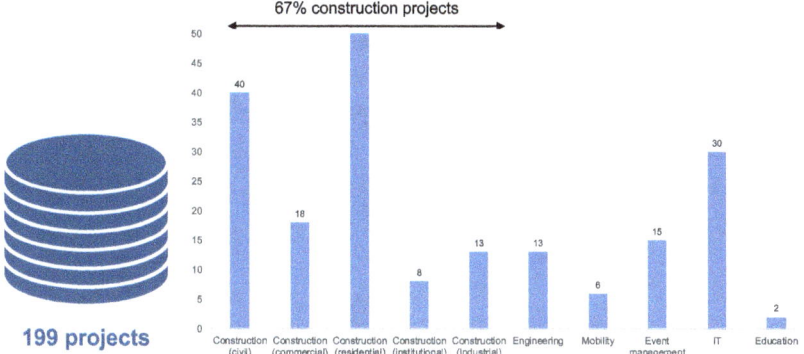

Fig. 9.3 Empirical database (2024)

but I nevertheless want to emphasize once again that they offer many advantages that empirical projects do not have.

The most significant advantage is the complete control that they offer. Since artificial projects need to be generated, their parameters can be much better controlled, making it easier to gain general insights in research studies. In the first part of this book, we continuously demonstrated that both network indicators and resource indicators have a strong impact on the complexity of the scheduling problem, insights that we likely would not have obtained if we had only looked at empirical projects.

That does not mean I want to undermine the benefits of empirical projects. There are a plenty of arguments as to why empirical projects are often better than artificial projects, with the lack of reality being the main reason why artificial projects sometimes fall short. But even with artificial projects, efforts can be made to enhance realism. Never underestimate the creativity and imagination of researchers, especially when they are in need of data for their research.

Our research into the resource-constrained project scheduling problem in the first part of this book also led to insights and ideas for expanding this problem, along with its associated artificial data, to make it more realistic. Generating realistic extensions of this scheduling problem is relatively straightforward, but delving into these extensions required us to generate new project data once more. And since the collection of empirical projects is a time-consuming and often challenging process, we once again decided to rely on artificial projects for this reason. Some of these extensions, along with their corresponding data, are discussed in the next and final chapter of this book.

If you truly believed that we already had enough project data, then you will discover that there is so much more to explore.

10

Do We Have Enough Project Data? (Part 2)

I admit that this chapter is the least enjoyable chapter of this book. It is not even meant to be. It is not written to sit back and read a story. While I do hope that the previous chapters were enjoyable to read, this final chapter primarily serves as a summary, supplemented with references to research studies for those seeking further details. The purpose of this chapter is to provide an overview of the available artificial project data that we have generated and analyzed in other research studies. Many of these data were extensively discussed in the first part of this book, but many others are new and have not yet been covered.

To structure this overview of our additional artificial project data, I must first return to the basic assumptions of the resource-constrained project scheduling problem, something I already did in the introduction of this book when I first mentioned the problem. The core problem is very strictly defined so that researchers can work on the same problem, fostering interaction and competitiveness, and accelerating new results.

In the basic version of the RCPSP, a project is described as follows:

A project consists of a series of activities that are interconnected by sequence dependencies, which are modeled as a network. Each activity requires a specific number of resources from a limited pool of resources available.

In other words, the characteristics of the core problem consist of nothing more than a *network* and *resources*, and that was clearly visible in the first part of this book. We were generating and manipulating artificial projects there, making them easier at some times, or more difficult at other times, by modifying both the network and resource indicators. Other changes were simply not possible because the projects themselves consisted of nothing more.

However, the basic version of the RCPSP can be extended in various ways, which are classified into three categories below:

- The application of the network and the precise combinations of activity duration and resources are strictly enforced, with minimal room for additional flexibility.
- Resources are treated as anonymous entities, devoid of human characteristics such as individual shortcomings and strengths.
- Projects operate in isolation, not integrated into a broader portfolio, thereby lacking synergy between these different initiatives.

Each extension naturally entails additional data for the projects, and therefore additional parameters to describe these new project characteristics. In the following paragraphs, I will provide a brief overview of the new artificial project data, leaving the details of the new parameters to the reader who is willing to read our published papers, which I will also reference each time.

10.1 Resources

The artificial projects generated for the basic version of the scheduling problem with resources were extensively discussed in the first part of the book, so it is unnecessary to discuss them again in detail. Therefore, I only provide a brief overview here with references to the studies that were also used in previous chapters. All this data was eventually compiled under the RCPLIB set, an abbreviation for resource-constrained project library. This library consists of a total of 13 different sets of artificial projects, nine of which were previously discussed in part I (Side A) of this book, while four have not been addressed at all. A brief overview is provided below:

Set 1 The PSPLIB instances were discussed in the prologue of Part I of this book and come from the following study:

> Kolisch, R. and Sprecher, A. (1996). PSPLIB - A project scheduling problem library. *European Journal of Operational Research*, 96:205–216.

Sets 2–6 A summary of five datasets (RG30, RG300, DC1, DC2, and MT) is introduced in Chaps. 1 and 2 and summarized in the following study:

> Vanhoucke, M., Coelho, J., and Batselier, J. (2016). An overview of project data for integrated project management and control. *Journal of Modern Project Management*, 3(2):6–21.

Set 7 The full NetRes dataset can be generated using the MT set (only network data) and the NetRes data files (only resource data) containing almost 4 million projects as discussed in Chap. 3. The 1kNetRes is a subset of NetRes and contains 3810 instances. A reference can be found in the following study:

> Vanhoucke, M. and Coelho, J. (2018). A tool to test and validate algorithms for the resource-constrained project scheduling problem. *Computers and Industrial Engineering*, 118:251–265.

Set 8 A dataset CV with 623 hard instances for exact procedures was the subject of Chap. 5.

> Coelho, J. and Vanhoucke, M. (2020). Going to the core of hard resource-constrained project scheduling instances. *Computers and Operations Research*, 121:104976.

Set 9 A dataset sD with 390 hard instances for heuristic procedures was the subject of Chap. 7.

> Coelho, J. and Vanhoucke, M. (2023). New resource-constrained project scheduling instances for testing (meta-)heuristic scheduling algorithms. *Computers and Operations Research*, 153:106165.

The attentive reader will, of course, recall that each instance consists of four versions. The original instances are those discussed above, but for most of them, lowRU and highRD versions are also available, as proposed in Chap. 6. Furthermore, most of these original instances also have an optimal equivalent version, which we referred to as the keepOPT version in Chap. 8.

In addition to these artificial project datasets, additional datasets were generated in a number of studies that were not covered in this book (but are available on our data website). A brief overview is provided below.

Set 10 A new dataset MH with 461 instances for meta-heuristic procedures, solving the scheduling problem through a combination of exact and heuristic methods, is also available. This set is presented in the following study:

> Vanhoucke, M., & Coelho, J. (2024a). A matheuristic for the resource-constrained project scheduling problem. *European Journal of Operational Research*, 319:711–725.

Set 11 The VNR set is another subset of NetRes with 1750 instances and was designed to introduce variability between the projects. VNR stands for variability in the NetRes set. The set is proposed in the following study:

> Van Eynde, R., Vanhoucke, M., & Coelho, J. (2024). On the summary measures for the resource-constrained project scheduling problem. *Annals of Operations Research*, 337:593–625.

Sets 12 and 13 The LPP and LPSP datasets are large-sized instance. Most projects from the previous sets were not large enough, prompting us to generate new and larger projects. The LP stands for large projects, with an additional P in LPP to indicate that the set mainly consists of parallel projects. The SP in LPSP is an abbreviation indicating that this set consists of both serial and parallel networks. Both sets are proposed in the following study:

> Luo, J., Vanhoucke, M., Coelho, J., and Guo, W. (2022). An efficient genetic programming approach to design priority rules for resource-constrained project scheduling problem. *Expert Systems with Applications*, 198:116753.

There is a lot of data available for this challenging project scheduling problem. You might think that with the many powerful algorithms and the abundance of data, this problem would be nearly solved by now. Fortunately, despite significant progress, there is still much work to be done, and for many project instances, the problem is far from being solved to optimality. That is fortunate for the many researchers who enjoy tackling this problem.

10.2 Flexibility

The strict network structure in which activities always have a fixed place in the network with a fixed duration and a fixed portion of resource use is often too rigid to construct realistic project schedules. In some cases, much more flexibility is needed, and this can be introduced in various ways. Together with Vincent Van Peteghem and Tom Servranckx, we incorporated two forms of additional flexibility in the network structure, and for each of them, we generated an additional artificial dataset.

Multiple Activity Modes The fixed durations and resource usage for activities of the RCPSP were expanded to a flexible usage where both the duration and resource use can vary. A realistic application of this is to express the size of the activity in person days, where 4 days with

3 resources are equivalent to 3 days with 4 resources. This gives the scheduling algorithm extra freedom to adjust the durations and resource requirements, as long as the total work content in person days remains the same. However, it is not always easy to calculate this work content as the product of duration and resource usage, as there are often nonlinear relationships that determine the work content. While it is easy to accept that more resources equate to less time, it is not as straightforward to accept that if one woman can deliver a baby in nine months, it also means that nine women can deliver a baby in one month. This is why the so-called multimode resource-constrained project scheduling problem incorporates different nonlinear possibilities to model the activity duration and resource usage. The extension leads to multiple modes for each activity from which a choice must be made before a project schedule can be constructed.

We have generated additional artificial project data for this extension as well, which we have compiled in the MMLIB set, an abbreviation for the multimode library. In the study below, we presented three different datasets for the extended scheduling problem, known as the multimode libraries MMLIB50, MMLIB100, and MMLIB+:

> Van Peteghem, V. and Vanhoucke, M. (2014). An experimental investigation of metaheuristics for the multi-mode resource-constrained project scheduling problem on new dataset instances. *European Journal of Operational Research*, 235:62–72.

Alternative Subgraphs The flexibility can be further expanded by not focusing on activity/resource modes, but by building flexibility into the network itself. It is often the case that alternative possibilities are incorporated into the network, and the project manager must make a choice during schedule construction as to which alternative is preferable. For example, it is often the case that alternative technologies can be used in a project, each with their own advantages and disadvantages, and each alternative represents a different part of the network within the project's complete network. Such a subnetwork is called an alternatives subgraph. We have generated quite a bit of data for this so-called resource-constrained project scheduling problem with alternative subgraphs, and we have compiled everything into the ASLIB set (consisting of subsets 0 to 6), an abbreviation for the alternative subgraph library.

The ASLIB0 contains two separate sets of artificial projects proposed in:

> Servranckx, T. and Vanhoucke, M. (2019). A tabu search procedure for the resource-constrained project scheduling problem with alternative subgraphs. *European Journal of Operational Research*, 273:841–860.

The project data of the ASLIB0 dataset contains various relationships between the alternative subgraphs in the network. However, some years later, we enhanced them with richer features to further increase realism. That is why we generated five new datasets labeled as ASLIB1 to ASLIB5, for which a summary is presented in:

> Servranckx, T. and Vanhoucke, M. (2023). New datasets for the resource-constrained project scheduling problem with alternative subgraphs. *Working paper (under submission)*.

Last but not least, we were impressed by the practical relevance of this extension, which gave us a reason to collect a set of empirical project data proposed as ASLIB6 in:

> Servranckx, T., Vanhoucke, M., and Vanhouwaert, G. (2020). Analysing the impact of alternative network structures on resource-constrained schedules: Artificial and empirical experiments. *Computers and Industrial Engineering*, 148:106706.

10.3 People Skills

In the basic version of the scheduling problem, resources are considered as anonymous entities, and activities use a number of them without distinguishing their performance and skills. This is also the case in most software tools but is, of course, a simplification of reality. Resources are not anonymous entities but usually consist of individuals with a heart, a soul, and their own personality. And people can be so different from each other that it is important to take that into account during the construction of a project schedule.

In the so-called multiskilled resource-constrained project scheduling problem, the demand of activities for anonymous resources is replaced by a demand for skills. Such an approach fundamentally changes the scheduling problem because now the right individuals with the right skills must be assigned to those activities before a project schedule can be created. The algorithms for such a problem thus consist of a resource assignment procedure, which does not occur in the basic version of the problem, and a scheduling procedure, making it even more complex than the already high complexity of the basic problem. The total number of resources is still limited but now consists of individuals with skills.

We have generated artificial projects in various ways to address this issue, and we have compiled them into the so-called MSLIB, an abbreviation for the multiskilled library. The problem has been investigated in the literature under different settings, and four new datasets, MSLIB1 to MSLIB4, with artificial projects to test existing and new procedures are proposed. Furthermore, a fifth dataset, MSLIB5, with empirical projects is also included. These five datasets are proposed in the following study co-authored by Jakob Snauwaert:

Snauwaert, J. and Vanhoucke, M. (2023). A classification and new benchmark instances for the multi-skilled resource-constrained project scheduling problem. *European Journal of Operational Research*, 307:1–19.

It is a widely accepted truth that individuals are intricate beings, each with their distinct set of strengths and weaknesses. This holds true not only in everyday life but also in the realm of project management. It should come as no surprise, then, that we have also generated data for this purpose.

10.4 Project Portfolios

Scheduling a single project with scarce resources is a complex endeavor, requiring careful coordination of activities, resources, and timelines to avoid resource conflicts and ensure a minimal project duration. However, constructing the schedule for multiple projects within a portfolio adds another layer of complexity. This is because projects are often interconnected, with dependencies and shared resources between them. Coordinating the timing and allocation of resources across multiple projects becomes a challenging puzzle, requiring extended algorithms for effective resource management to optimize the overall portfolio objective. While the single RCPSP presents its challenges, the complexity escalates significantly when managing multiple projects within a portfolio.

But we are precisely interested in such a challenge, and therefore, we have also generated artificial project portfolio data for the so-called resource-constrained multi-project scheduling problem. The generation process to construct such multi-project data libraries is fundamentally different from the generation mechanism used for single-project data generation, and we have compiled our new data into the MPLIB set, an abbreviation for multi-project library. This extended generation process is discussed in two published papers. In the first study, the first multi-project library (MPLIB1) is presented and compared with the existing project portfolio data from the literature.

> Van Eynde, R. and Vanhoucke, M. (2020). Resource-constrained multi-project scheduling: Benchmark datasets and decoupled scheduling. *Journal of Scheduling*, 23:301–325.

In a follow-up paper, a second multi-project library (MPLIB2) is presented that better reflects the true characteristics of project portfolios, which is published in the following study:

> Van Eynde, R. and Vanhoucke, M. (2022). New summary measures and datasets for the multi-project scheduling. *European Journal of Operational Research*, 299:853–868.

10.5 Where Is Lisbon?

I wrote at the beginning of this chapter that it would not be a chapter where an interesting story would be told, as was the case in the first part of this book.

I kept my promise.

But that does not mean that behind this chapter, there is no story. It may not be a research story centered around Lisbon, as the previous chapters were, but each study of this chapter conceals a research narrative too. It is just that this time, I did not delve too deeply into the story behind the data.

But I still hope that amidst the many new artificial project databases presented here, you also got to know the people behind the story a little bit. Just like the journey that I took with José in the first part of this book, this story, although very briefly told, is a tale of tackling challenges that led to new questions and even more challenges. And just like in any other research study, I did not make the journey on my own.

I do not like to travel alone, actually.

Academic research is and remains a journey into the unknown, and I am glad to undertake this journey with several fantastic people from my Operations Research & Scheduling group. It is often a pleasant journey, but just as frequently a difficult one, yet look where it brought us...

It gave us a wide range of data for projects with scarce resources.

That in itself made the journey already worth the effort.

Afterword

Congratulations! You have reached the end of this book. It marks the culmination of a data-discovery journey in search of projects with scarce resources. A journey told from our personal research perspective.

This journey began in Valencia, Spain, where I met José Coelho, and led me to Lisbon, where I now live part time. I would like to thank him for co-authoring Chaps. 1 to 8. Without him, this book could never have been written.

I also want to express gratitude to the other members of my team for co-authoring Chaps. 9 and 10. Special thanks to Gaëtane Beernaert, my lovely wife, Tom Servranckx, my work partner in crime, and Sokko, my best friend, for proofreading every part of this book meticulously.

I would also like to thank two people in particular who have indirectly provided a lot of help. I believe they are too often not recognized enough for their excellent work. I am referring to Francisco Saldanha da Gama and Roman Słowiński, the editors of the two journals where I have published most of my work. It is often forgotten that the quality of journals depends not only on the quality of the papers, but also, and especially, on the editors. They choose the reviewers, interpret the reports with comments, and ultimately propose a final conclusion —

strict but fair. And even though that conclusion is usually very heavy (major revision, or even sometimes rejection), these editors often give us a chance to take the comments to heart and demonstrate that our research is valuable. This whole process of submitting and revising is often a lengthy and difficult one, but they consistently manage it with an unwavering dedication to steer everything in the right direction. Therefore, a special thanks to them. We should do this more often.

Additionally, I wish to thank my readers for following along with the narrative of this journey, and the professional project managers, especially those who have generously responded to my calls to share their data. I truly hope that we can sustain this exchange of ideas and data, bridging the gap between academia and practice, to continuously advance our knowledge in this wonderful research domain.

I hope this book becomes outdated very quickly.
I hope that this story helps create new ideas.
I hope it will be the start of much more research.

Why I Write Books

To end this book beautifully, I have included a summary of my previous project management books. I could do this under the guise of wanting to "stimulate further research" or wanting to "share my passion for the profession," but perhaps there is just a bit of a marketeer in me.

I write books for several reasons. As Salman Rushdie once said, "*a book is a version of the world. If you do not like it, ignore it, or offer your own version in return.*" While that may be the primary reason why I continue to write, there are additional motivations behind my work.

While I find my job incredibly challenging, academic research can sometimes be frustrating. It involves numerous discussions and extensive computer experiments in search of results, which are then submitted to journals with hopes of acceptance. Referees evaluate our papers and provide feedback, which we use to improve our research. Although this feedback often enhances our work, sometimes it steers it in directions that we had not intended. Or sometimes even in directions that we did not even want to go.

Even in academic research, compromises must be made.

Fortunately, when writing books, this is much less the case. A book is more of a personal story where you can share your narrative without major compromises. Everything I write is, of course, based on academic research, but there is more freedom in how it can be written. It allows room to sometimes write sentences that would never be accepted in an academic journal. I love that freedom, and it is the main reason why I keep writing.

Each of my books tells the project management story differently. Every narrative is grounded in academic research and aims to stimulate practical relevance. Each book is tailored to a different audience. Here is a brief summary of those published before:

- **Measuring Time: Improving Project Performance Using Earned Value Management**: Everything I have learned about Earned Value Management comes from my first book. I was not sure at that time why I wanted to write a book, but maybe winning my very first research award from the International Project Management Association encouraged me to continue.
- **Project Management with Dynamic Scheduling: Baseline Scheduling, Risk Analysis, and Project Control**: This book tells the story of my lectures at universities in Ghent, London, and Beijing. It is written as a student manual for masters' students who like having something tangible to refer to.
- **Integrated Project Management and Control: First Comes the Theory, then the Practice**: This book summarizes commonly used metrics in Earned Value Management and Schedule Risk Analysis, with performance reports and software tutorials. It is not for casual reading but essential if you want to implement the techniques yourself to manage your projects.
- **Integrated Project Management Sourcebook: A Technical Guide to Project Scheduling, Risk, and Control**: This book offers technical summaries in more than 70 separate articles, along with test questions and model answers. Most material is also available on our PM Knowledge Center website.

Afterword

- **The Data-Driven Project Manager: A Statistical Battle Against Project Obstacles**: This technical business novel is used in company trainings at Vlerick Business School and University College London. It presents methodologies as an integrated change process for implementing data-driven project management.
- **The Illusion of Control: Project Data, Computer Algorithms, and Human Intuition for Project Management and Control**: This book explores many academic research studies and presents 20 years of collaboration with academics and professionals. It covers themes like statistical project control, forecasting, and machine learning. The positive feedback on this book inspired me to write the book you are currently reading.

Six books already, and now the seventh. I hope you understand that I am proud of this, and I could not resist showing them all in a beautiful picture. Figure 1 not only shows that yellow is the favorite color of my publisher (Springer) but that much of the work of my research group has received recognition through awards from the International Project Management Association, the Project Management Institute, and the College of Performance Management.

Fig. 1 My project management bookstore

The Future of Project Data

I am uncertain about the future of data in project management. While I have read optimistic accounts about the potential impact of machine learning and artificial intelligence on project management, some skeptics doubt its transformative power. I am not sure if it will all happen as quickly as some hope or claim, but I am convinced that the focus on data, in any form, will become increasingly important for managing projects.

What I have definitely taken away from reading the brilliant book "*Thinking, Fast and Slow*" by late Daniel Kahneman is that he warns us against relying too heavily on our intuition due to our susceptibility to biases and errors. Data helps us put these biases into perspective, and reliable decisions cannot be made without it. A data-driven approach to project management is thus a way to overcome our inherent biases.

I do not know what the future of data will bring, but I remain hopeful and have no intention of slowing down my academic research. And José also has no plans to slow down. In a recent meeting, we came up with a number of ideas that reignited our enthusiasm once again. It seems there will not be much time to enjoy the Sun after all. The supercomputer is already running several tests. We might come up with a new algorithm that combines a mathematical model with a meta-heuristic to better solve our all-time favorite resource-constrained project scheduling problem (Fig. 2).

We are hopeful, and our preliminary results are promising, but much work still needs to be done.

It might become the journey for a next book.

Fig. 2 José (left) and Mario (right) in Ghent, Belgium (August 2023)

Research Background (*a bit tedious to read*)

I tried to use as few references to academic papers as possible in this book, except for those mentioned at the beginning of each chapter, which formed the basis of each chapter. Nevertheless, I have not succeeded in never referring to other studies, and I am therefore obliged to cite a number of references per chapter. These are listed below, with a brief explanation. I have also listed the references to the studies that formed the basis of each chapter again to make sure you have the full reference details.

Side A. The Quest for Project Data (Featuring José Coelho)

In this prologue, I made the very first reference to the well-known PSPLIB dataset, which can be found in the study by Kolisch and Sprecher (1996).

I also could not resist mentioning the brilliant and one of the most performant branch-and-bound procedures for solving the RCPSP, proposed in the study by Demeulemeester and Herroelen (1992).

Finally, I could not help but mention one of the very first datasets used in a lot of research in project scheduling. This so-called Patterson dataset was introduced in the study by Patterson (1976).

Chapter 1. Do We Have Enough Project Data? (Part 1)

This chapter is based on the study by Vanhoucke et al. (2008).

In this chapter, I also introduced the random network generator RanGen1 for the very first time, which formed the basis for developing the RanGen2 generator (for which the reference is provided at the beginning of the chapter). The reference for the RanGen1 generator is Demeulemeester et al. (2003).

I referred to a paper at the beginning of this chapter that has had a significant impact on my research. It is a paper by two of the greatest researchers in project scheduling, for whom I have a lot of respect. The study can be found in Elmaghraby and Herroelen (1980).

Last but not least, I referred to the "phase transitions" study in project scheduling, which can be found in Herroelen and De Reyck (1999).

Chapter 2. Do the Projects Exist in Reality?

This chapter is based on the two following studies:

- A summary of the current databases is given in the study by Vanhoucke et al. (2016).
- The empirical project database is presented in the study by Batselier and Vanhoucke (2015).

In this chapter, I mentioned many existing datasets, each originating from a different study. The references to those studies are as follows:

- The RG300 dataset comes from the study by Debels and Vanhoucke (2007).
- The DC1 dataset comes from the study by Vanhoucke et al. (2001).

- The DC2 dataset comes from the study by Vanhoucke (2010b).
- The MT dataset comes from the book by Vanhoucke (2010a).

Chapter 3. Do We Have Good Schedules for the Projects?

This chapter is based on the study by Vanhoucke and Coelho (2018).

Chapter 4. Can We Solve Every Project Instance?

This chapter is based on the study by Coelho and Vanhoucke (2018).

The 13 lower bounds used in our branch-and-bound procedure come from the study by Klein and Scholl (1999).

In this chapter, I also referred to the study with Weikang Guo, which can be found in Guo et al. (2023).

Finally, I also referred to the 12 existing branch-and-bound procedures (of which 10 were included in the composite branch-and-bound procedure). These can be found in the studies by Patterson and Huber (1974), Stinson et al. (1978), Talbot and Patterson (1978), Christofides et al. (1987), Bell and Park (1990), Demeulemeester and Herroelen (1992), Demeulemeester and Herroelen (1997), Mingozzi et al. (1998), Brucker et al. (1999), Nazareth et al. (1999), Dorndorf et al. (2000), and Sprecher (2000).

Chapter 5. Why Is the RCPSP So Difficult? (Part 1)

This chapter is based on the study by Coelho and Vanhoucke (2020).

156   Research Background (*a bit tedious to read*)

Chapter 6. Can We Make the Projects Easier? (Part 1)

This chapter is based on the study by Vanhoucke and Coelho (2021).

Chapter 7. Why Is the RCPSP So Difficult? (Part 2)

This chapter is based on the study by Coelho and Vanhoucke (2023).
   In this chapter, I used three meta-heuristics developed in previous studies. The references to those studies are as follows:

- The genetic algorithm (GA) comes from the study by Debels and Vanhoucke (2007).
- The scatter search (SS) comes from the study by Debels et al. (2006).
- The electromagnetic heuristic (EM) comes from the study by Debels and Vanhoucke (2006).

Chapter 8. Can We Make the Projects Easier? (Part 2)

This chapter is based on the study by Vanhoucke and Coelho (2024b).
   In this chapter, I referred to the study with Rob van Eynde, which can be found in Van Eynde and Vanhoucke (2022).

Side B. The Relevance of Project Data (Featuring Other Researchers)

Chapter 9. How to Make the Project Data Practically Relevant?

This chapter is based on the following three studies:

- The calibration procedure has been empirically validated by Colin and Vanhoucke (2016).
- The calibration procedure has been extended to human partitioning by Vanhoucke and Batselier (2019a).
- The calibration procedure has been extended to statistical partitioning by Vanhoucke and Batselier (2019b).

The original calibration procedure used as the foundation for Chapter 9 was initially proposed in the excellent study by Trietsch et al. (2012).

Parkinson's law is one of the foundations of the calibration methods in this chapter, and the original reference can be found in Parkinson (1957).

Chapter 10. Do We Have Enough Project Data? (Part 2)

In this chapter, I listed many references to studies in which new extended datasets were introduced. However, I mentioned them all in the text and will therefore not repeat them here.

References

Batselier, J., & Vanhoucke, M. (2015). Construction and evaluation framework for a real-life project database. *International Journal of Project Management, 33*, 697–710.

Bell, C. E., & Park, K. (1990). Solving resource-constrained project scheduling problems by a* search. *Naval Research Logistics (NRL), 37*(1), 61–84.

Brucker, P., Drexl, A., Möhring, R., Neumann, K., & Pesch, E. (1999). Resource-constrained project scheduling: notation, classification, models, and methods. *European Journal of Operational Research, 112*, 3–41.

Christofides, N., Alvarez-Valdes, R., & Tamarit, J. (1987). Project scheduling with resource constraints: A branch and bound approach. *European Journal of Operational Research, 29*(3), 262 – 273.

Coelho, J., & Vanhoucke, M. (2018). An exact composite lower bound strategy for the resource-constrained project scheduling problem. *Computers and Operations Research, 93*, 135–150.

Coelho, J., & Vanhoucke, M. (2020). Going to the core of hard resource-constrained project scheduling instances. *Computers and Operations Research, 121*, 104976.

Coelho, J., & Vanhoucke, M. (2023). New resource-constrained project scheduling instances for testing (meta-)heuristic scheduling algorithms. *Computers and Operations Research, 153*, 106165.

Colin, J., & Vanhoucke, M. (2016). Empirical perspective on activity durations for project management simulation studies. *Journal of Construction Engineering and Management, 142*(1), 04015047.

Debels, D., De Reyck, B., Leus, R., & Vanhoucke, M. (2006). A hybrid scatter search/electromagnetism meta-heuristic for project scheduling. *European Journal of Operational Research, 169*, 638–653.

Debels, D., & Vanhoucke, M. (2006). The electromagnetism meta-heuristic applied to the resource-constrained project scheduling problems. *Lecture Notes in Computer Science, 3871*, 259–270.

Debels, D., & Vanhoucke, M. (2007). A decomposition-based genetic algorithm for the resource-constrained project scheduling problems. *Operations Research, 55*, 457–469.

Demeulemeester, E., & Herroelen, W. (1992). A branch-and-bound procedure for the multiple resource-constrained project scheduling problem. *Management Science, 38*, 1803–1818.

Demeulemeester, E., & Herroelen, W. (1997). New benchmark results for the resource-constrained project scheduling problem. *Management Science, 43*, 1485–1492.

Demeulemeester, E., Vanhoucke, M., & Herroelen, W. (2003). RanGen: A random network generator for activity-on-the-node networks. *Journal of Scheduling, 6*, 17–38.

Dorndorf, U., Pesch, E., & Phan-Huy, T. (2000). A branch-and-bound algorithm for the resource-constrained project scheduling problem. *Mathematical Methods of Operations Research, 52*, 413–439.

Elmaghraby, S., & Herroelen, W. (1980). On the measurement of complexity in activity networks. *European Journal of Operational Research, 5*, 223–234.

Guo, W., Vanhoucke, M., & Coelho, J. (2023). A prediction model for ranking branch-and-bound procedures for the resource-constrained project scheduling problem. *European Journal of Operational Research, 306*, 579–595.

Herroelen, W., & De Reyck, B. (1999). Phase transitions in project scheduling. *Journal of the Operational Research Society, 50*, 148–156.

Klein, R., & Scholl, A. (1999). Computing lower bounds by destructive improvement: An application to resource-constrained project scheduling. *European Journal of Operational Research, 112*, 322–346.

Kolisch, R., & Sprecher, A. (1996). PSPLIB—a project scheduling problem library. *European Journal of Operational Research, 96*, 205–216.

Luo, J., Vanhoucke, M., Coelho, J., & Guo, W. (2022). An efficient genetic programming approach to design priority rules for resource-constrained project scheduling problem. *Expert Systems with Applications, 198*, 116753.

Mingozzi, A., Maniezzo, V., Ricciardelli, S., & Bianco, L. (1998). An exact algorithm for the resource constrained project scheduling problem based on a new mathematical formulation. *Management Science, 44*, 714–729.

Nazareth, T., Verma, S., Bhattacharya, S., & Bagchi, A. (1999). The multiple resource constrained project scheduling problem: A breadth-first approach. *European Journal of Operational Research, 112*(2), 347–366.

Parkinson, C. N. (1957). *Parkinson's law and other studies in administration.* Houghton Mifflin.

Patterson, J. (1976). Project scheduling: The effects of problem structure on heuristic scheduling. *Naval Research Logistics, 23*, 95–123.

Patterson, J. H., & Huber, W. (1974). A horizon-varying, zero-one approach to project scheduling. *Management Science, 20*, 990–998.

Servranckx, T., & Vanhoucke, M. (2019). A tabu search procedure for the resource-constrained project scheduling problem with alternative subgraphs. *European Journal of Operational Research, 273*, 841–860.

Servranckx, T., & Vanhoucke, M. (2023). New datasets for the resource-constrained project scheduling problem with alternative subgraphs. Working paper (under submission).

Servranckx, T., Vanhoucke, M., & Vanhouwaert, G. (2020). Analysing the impact of alternative network structures on resource-constrained schedules: Artificial and empirical experiments. *Computers and Industrial Engineering, 148*, 106706.

Snauwaert, J., & Vanhoucke, M. (2023). A classification and new benchmark instances for the multi-skilled resource-constrained project scheduling problem. *European Journal of Operational Research, 307*, 1–19.

Sprecher, A. (2000). Scheduling resource-constrained projects competitively at modest memory requirements. *Management Science, 46*, 710–723.

Stinson, J., Davis, E., & Khumawala, B. (1978). Multiple resource-constrained scheduling using branch-and-bound. *IIE Transactions, 10*, 252–259.

Talbot, F., & Patterson, J. (1978). An efficient integer programming algorithm with network cuts for solving resource-constrained scheduling problems. *Management Science, 24*, 1163–1174.

Trietsch, D., Mazmanyan, L., Govergyan, L., & Baker, K. R. (2012). Modeling activity times by the Parkinson distribution with a lognormal core: Theory and validation. *European Journal of Operational Research, 216*, 386–396.

Van Eynde, R., & Vanhoucke, M. (2020). Resource-constrained multi-project scheduling: Benchmark datasets and decoupled scheduling. *Journal of Scheduling, 23*, 301–325.

Van Eynde, R., & Vanhoucke, M. (2022). New summary measures and datasets for the multi-project scheduling. *European Journal of Operational Research, 299*, 853–868.

Van Eynde, R., Vanhoucke, M., & Coelho, J. (2024). On the summary measures for the resource-constrained project scheduling problem. *Annals of Operations Research, 337*, 593–625.

Van Peteghem, V., & Vanhoucke, M. (2014). An experimental investigation of metaheuristics for the multi-mode resource-constrained project scheduling problem on new dataset instances. *European Journal of Operational Research, 235*, 62–72.

Vanhoucke, M. (2010a). *Measuring time—improving project performance using earned value management*, volume 136 of International Series in Operations Research and Management Science. Springer.

Vanhoucke, M. (2010b). A scatter search heuristic for maximising the net present value of a resource-constrained project with fixed activity cash flow. *International Journal of Production Research, 48*, 1983–2001.

Vanhoucke, M., & Batselier, J. (2019a). Fitting activity distributions using human partitioning and statistical calibration. *Computers and Industrial Engineering, 129*, 126–135.

Vanhoucke, M., & Batselier, J. (2019b). A statistical method for estimating activity uncertainty parameters to improve project forecasting. *Entropy, 21*, 952.

Vanhoucke, M., & Coelho, J. (2018). A tool to test and validate algorithms for the resource-constrained project scheduling problem. *Computers and Industrial Engineering, 118*, 251–265.

Vanhoucke, M., & Coelho, J. (2021). An analysis of network and resource indicators for resource-constrained project scheduling problem instances. *Computers and Operations Research, 132*, 105260.

Vanhoucke, M., & Coelho, J. (2024a). A matheuristic for the resource-constrained project scheduling problem. *European Journal of Operational Research, 319*, 711–725.

Vanhoucke, M., & Coelho, J. (2024b). Reducing the feasible solution space of resource-constrained project instances. *Computers and Operations Research, 165*, 106567.

Vanhoucke, M., Coelho, J., & Batselier, J. (2016). An overview of project data for integrated project management and control. *Journal of Modern Project Management, 3*(2), 6–21.

Vanhoucke, M., Coelho, J., Debels, D., Maenhout, B., & Tavares, L. (2008). An evaluation of the adequacy of project network generators with systematically sampled networks. *European Journal of Operational Research, 187*, 511–524.

Vanhoucke, M., Demeulemeester, E., & Herroelen, W. (2001). On maximizing the net present value of a project under renewable resource constraints. *Management Science, 47*, 1113–1121.

GPSR Compliance

The European Union's (EU) General Product Safety Regulation (GPSR) is a set of rules that requires consumer products to be safe and our obligations to ensure this.

If you have any concerns about our products, you can contact us on

ProductSafety@springernature.com

In case Publisher is established outside the EU, the EU authorized representative is:

Springer Nature Customer Service Center GmbH
Europaplatz 3
69115 Heidelberg, Germany

www.ingramcontent.com/pod-product-compliance
Lightning Source LLC
LaVergne TN
LVHW010959250326
834688LV00003B/15